Energy Conservation in Industry –
Combustion, Heat Recovery and Rankine Cycle Machines

Energy R&D Programme of the
Commission of the European Communities

Energy Conservation in Industry –
Combustion, Heat Recovery and Rankine Cycle Machines

Proceedings of the Contractors' Meetings held in Brussels on 10 and 18 June, and 29 October 1982

Edited by

H. Ehringer, G. Hoyaux and P. A. Pilavachi

Directorate-General for Science, Research and Development, Commission of the European Communities

Springer-Science+Business Media, B.V.

Library of Congress Cataloging in Publication Data
Main entry under title:

Energy conservation in industry.

At head of title: Energy R&D programme of the Commission of the European Communities.
Meeting organized by the Directorate-General for Science, Research and Development, Commission of the European Communities.
1. Industry–Energy conservation–Congresses. 2. Combustion engineering–Congresses. 3. Heat recovery–Congresses. I. Ehringer, H. II. Hoyaux, G. III. Pilavachi, P. (Petros), 1944–
IV. Commission of the European Communities. Directorate-General for Science, Research, and Development.
TJ163.27.E49 1983 621.402 83-3150
ISBN 978-90-481-8382-1 ISBN 978-94-015-7674-1 (eBook)
DOI 10.1007/978-94-015-7674-1

Organization of the Contractors' meetings by
Commission of the European Communities
Directorate-General Science, Research and Development, Brussels

Publication arrangements by
Commission of the European Communities
Directorate-General Information Market and Innovation, Luxembourg

EUR 8434

Originally published by ECSC, EEC, EAEC, Brussels and Luxembourg in 1983
Softcover reprint of the hardcover 1st edition 1983

PREFACE

The main objectives of the Community's energy policy consist of securing a sufficient energy supply for the present and future demand of its Member States and in reducing the Community's dependence on imported energy through the more rational use of energy and a broader diversification of supply. This requires a full set of common efforts at all levels, including energy research and development.

In the framework of its strategy for scientific and technological research the Commission of the European Communities has launched several R & D programmes in the field of energy. Theses programmes are implemented either directly in its Joint Research Centre or by concluding contracts with research institutions in the EC Member countries.

One of the most important short and medium term objectives of the present four-year Energy R & D Programme (1979 - 1983) approved by the Council of Ministers on 13 September 1979 aims at fostering energy conservation technologies in the three main energy consuming sectors: domestic, industry and transport.

In the European Community more than 35% of the energy is consumed in industry. About 23% of the imported oil is used in this sector. Two thirds of this energy are used for heating purposes. Consequently, there is a high potential for residuary heat recovery.

In order to enhance rational use of energy and energy conservation in the Member States of the Community, it is necessary, among other things, to promote the development of combustion and energy recovery techniques.

The Commission has therefore concluded with industries, public organisations and universities of the EC Member States a number of R & D contracts dealing with combustion and heat recovery techniques. The progress achieved in these contracts was reported in contractors meetings, which took place on 10 and 12 June 1982 in Brussels. There was also a meeting on 29 October 1982 on energy cascading and ORC engines.

The papers presented at these meetings and the conclusions of the discussions are collected in the present book, which is intended to inform the Member States, research bodies and other interested parties about existing R & D work in this area.

I should like to address may sincere thanks to all the authors of papers and speakers of these meetings. I should also thank the chairmen of these meetings, Mr. David REAY and Dr. Giovanni LIMIDO, who animated the discussion with comments and questions. Last but not least, I extend my thanks to the editors of these proceedings, who summarized the content of the meetings and the discussion.

May I express the hope that the work reported will contribute to rapid progress in the development of practical application of improved combustion and heat recovery techniques, thereby contributing to rational use of energy throughout the European Community.

Dr. A. STRUB
Director
Head of the Energy R & D Programme

ATTENDANCE LIST

1. CONTRACTORS

SESSION I : COMBUSTION

Dr. V. SUBRAMANIAN	THE BRITISH ALUMINIUM COMPANY
R.L. McGLASHAN	THE BRITISH ALUMINIUM COMPANY
J.L. ROTH	IRSID
W.E. WHITMAN	LEATHERHEAD FOOD RESEARCH ASSOCIATION

SESSION II : HEAT EXCHANGERS - HEAT RECOVERY

S. GALANT	BERTIN
A. ROJEY	INSTITUT FRANCAIS DU PETROLE
A. GREHIER	INSTITUT FRANCAIS DU PETROLE
J. DREUILHE	CREUSOT-LOIRE
B. CHLIQUE	CREUSOT-LOIRE
M. MISCHIATTI	FBM
C. ANTONINI	FBM
A. CALABRO	TEKSID
D. McCALLUM	JAMES HOWDEN AND CO.
G. CHAFFEY	UKAEA Harwell
J.C. RALPH	UKAEA Harwell
A.D. ROBERTSON	BRITISH STEEL CORPORATION
Dr. R. THORNTON	COURTAULDS
M.C. QUIGLEY	COURTAULDS
M. GROLL	IKE

SESSION III : ENERGY CASCADING AND ORC ENGINES

	H. HOPMANN	MBB
	G. HUPPMANN	MBB
Prof.	E. MACCHI	TURBODEN
	M. GAIA	TURBODEN
	M. FOUCAULT	EDF
	S. GOLDSTEIN	CEA
	B. VRILLON	CEA
Dr.	P.W. O'CALLAGHAN	CRANFIELD INSTITUTE
	M.A. BELL	CRANFIELD INSTITUTE
	MOHEY HUSSEIN	CRANFIELD INSTITUTE
	R.J. WOOD	CRANFIELD INSTITUTE

2. COMMISSION

Dr.	H. EHRINGER	DG XII
	G. HOYAUX	DG XII
Dr.	P.A. PILAVACHI	DG XII
	P. ZEGERS	DG XII

3. CHAIRMEN

Dr.	J. LIMIDO	expert	IFP
	D.A. REAY		IRD

CONTENTS

INTRODUCTION

by

Hermann EHRINGER

Commission of the European Communities

Industry consumes 35% of the final energy in the European Community. About 23% of the imported oil is used in this area. Two thirds of this energy are intended for heating purposes. In spite of intensive rationalization efforts undertaken by industry in the last decade, there is still an important energy saving potential, in particular, as regards heat recovery and combustion techniques. Therefore, the European Community supports R & D work on energy saving by means of energy cascading techniques, heat exchangers and Rankine Cycle engines.

The purpose of the contractors meetings is to give the project leaders the opportunity to report on the progress of their work and on the results achieved. Moreover, these meetings serve to establish a link between contractors working in the same area and to strengthen cooperation on a European level.

The papers presented at these meetings and the discussion will be recorded in proceedings which will be published in order to meet the provisions laid down in the Council regulation EEC/2380/74 of 17 September 1974.

SESSION I - COMBUSTION

Chairman: D.A. REAY

SUMMARY OF SESSION I

COMBUSTION

D.A. Reay*), P.A. Pilavachi**)

There are many changes and improvements which can be made to combustion systems to raise their efficiency, and hence conserve energy. Improvements to existing systems can readily be made by "good housekeeping", manual control, and the application of the results of combustion product analysis. Old burners can be replaced by new systems, possibly of the self-recuperative type. A fully automated control system may be adopted, possibly in conjunction with changes in burner type and plant operation.

The 3 projects on combustion, describing activities within the Commission R & D Programme, cover all the aspects mentioned above. BRITISH ALUMINIUM is involved in a substantial programme of modifications to aluminium melting and holding furnaces which has already results in energy savings of 6-7 % on one installation. INSTITUT DE RECHERCHES DE LA SIDERURGIE FRANCAISE (IRSID) is attacking the fuel efficiency problem in another way, by modifying the heat exchangers used to preheat combustion air so that they operate at a higher efficiency. The third project, involving the LEATHERHEAD FOOD RESEARCH ASSOCIATION and a factory in London, centres around the use of a low cost combustion control system on boilers where variable steam demand exists.
(It should be noted that combustion features in several other sectors of the Community Energy R & D Programme, including fluidized beds and heat exchangers/heat recovery).

Efficiency of reverbatory furnaces - The project being carried out by BRITISH ALUMINIUM was established with the initial aim of being able to develop systems and procedures which would save 15 per cent of the energy used in the company's melting furnaces. To date, the work suggests that double this amount will be saved. This has been made possible by three principal measures :

(i) Improved burner specifications, particularly increased turn-down ratio

(ii) Application of instruments

(iii) Automated single-loop control of individual parameters.

Work is in hand which will lead to the implementation of a fully automated multi-loop control system, which will be applied to combustion control. Estimated savings on the three production furnaces involved in the development to date are £ 70,000 per annum (126.000 ECU) and a payback period of 6-8 months is anticipated.

---

*) International Research and Development, UK

**) Commission of the European Communities

Energetic optimization of furnaces and heat exchangers - The work at IRSID was not reported at the contractors'meeting. The project has as its aim the saving of energy in reheat furnaces used in the manufacture of steel strip and plates. Because of the high exhaust gas temperatures, the use of radiation recuperators is necessary. These can be of low efficiency, and it is hoped to improve the performance of systems, and also to study the effects of downstream convective heat exchangers. An improved heat recovery system for combustion air preheating should, it is estimated, lead to fuel savings of up to 10 per cent on each furnace.

Improved boiler control in food factories - LEATHERHEAD FOOD RESEARCH ASSOCIATION, in collaboration with PLGROUP (FOODS) LTD is studying cost-effective management of steam supply and distribution in a food factory. This follows on from a broader assessment of energy conservation potential in the food industry completed in the 1st Energy R & D Programme. The industrial partner is active in edible oil processing (see also "Food Industry") and employs three boilers supplying five major steam users and a number of less important ones.
It is intended to collect data on the magnitude and timing of the various steam demands. These data will form, together with information on boiler efficiencies at different ratings, the basis of the input to a micro-computer which will aid assessment of a boiler control strategy.
This project is in its early stages, but the control strategy has been formulated, and awaits quantitative data on steam demand.

EFFICIENCY OF REVERBERATORY FURNACES

Authors	:	V. SUBRAMANIAN
Contract number	:	EEB-1-103-UK
Duration	:	36 months 1 July 1980 - 30 June 1983
Head of project	:	Dr V. Subramanian
Contractor	:	EEC/BA
Address	:	The British Aluminium Company plc Technical Department Chalfont Park GERRARDS CROSS, Bucks. SL9 0QB

Summary

This paper describes the work carried out in the three stage (instrumented manual, single loop and fully automated microprocessor based control) programme aimed at achieving substantial fuel savings in heavy fuel oil and natural gas fired reverberatory melting and holding furnaces. The programme involves pre- and post-installation monitoring on four selected furnaces. New burners with improved fuel air supply systems, instrumentation and control equipment have been installed primarily for the control of total air/fuel flow, air/fuel ratio and furnace pressure. A selection of instruments and control equipment have been used.

Stage I has been completed. Stage II (single-loop control) is nearly complete and Stage III is in progress. Problems experienced with instruments and control equipment have been reported and preferred equipment has been identified. Use of burners with a high 'on-ratio' turndown and an air damper system has been a major factor in achieving fuel savings. Automatic control of air/fuel ratio using oxygen analysis is considered unsatisfactory.

Savings achieved have been up to:

- 67% on an experimental furnace.
- 45% on factory holding furnaces.
- 29% on a factory melting furnace.
- up to 50% on melting for individual melts.

The estimated savings over the three production furnaces is about £70,000 p.a.

1.1. Introduction

In B.A. there are over 50 reverberatory melting and holding furnaces which consume fuel (gas or oil) to the value of about £4 million per annum at 1980 costs. The furnace efficiencies in B.A. (generally typical of the whole industry) range from 10-25% and these low efficiencies are the result of operating practices, inadequate combustion and furnace control, and heat losses through the stack and the furnace structure. By making improvements in the above areas, i.e. by application of instrumentation and automatic furnace control systems, it is aimed to achieve a fuel saving of 15%.

In general, the end objective of the work is to achieve fuel savings in reverberatory melting and holding furnaces by developing a cost effective, fully automated, multi-loop combustion and furnace control system.

The programme of work over the three year contract period involves three overlapping stages of development on four selected furnaces (shown schematically in Fig. 1).

The stages of development are:

Stage I Instrumented manual control.

Stage II Development of single-loop automated control of individual parameters.

Stage III Fully automated, multi-loop control.

Four conventional reverberatory furnaces, for aluminium melting and holding, covering the different types of fuel used, have been selected for the work:

(i) A natural gas-fired melter (experimental) of 8 tonne capacity, at the Technical Department, Chalfont Park.

(ii) A heavy fuel oil-fired melter (No. 3), of 25 tonne capacity, at B.A. Alloys, Latchford.

(iii) A heavy fuel oil-fired holder (No. 3), of 20 tonne capacity, at B.A. Alloys, Latchford.

(iv) A natural gas-fired holder (No. 11) of 23 tonne capacity, at B.A. Rolled Products, Falkirk.

This paper covers work up to December 1981 with the following specific objectives:

- completion of preliminary surveys on the selected furnaces to establish baselines for fuel savings.
- modification of furnace and burners.
- application and assessment for Stage I instrumentation and Stage II control equipment.
- assessment of the suitability and reliability of the various instruments and equipment.
- assessment and specifications of equipment requirements for Stage III fully automated multi-loop control.
- evaluation of fuel savings.

Stage I has been completed.

Stage II is nearly completed and Stage III is in progress.

1.2 Description

The evaluation programme consists of the following:

1.1.1 Pre-installation monitoring

These were conducted on the four selected furnaces to establish base line data for fuel savings and to make specifications regarding furnace and burner modifications

1.2.2 Post-installation monitoring

Prior to post-installation monitoring the following work had to be carried out:

(a) Modifications to the furnaces and burners

All furnaces were modified in the following ways:

- new burners, where necessary, with improved fuel/air supply systems to give increased turn-down ratio and to achieve stoichiometric combustion.
- installation of air dampers in the flue to reduce the air ingress into the furnace and loss of heat in flue gases.
- location of various sensors in each furnace and flue.
- improved furnace insulation, where possible, to minimise structural heat loss.

The positions of the sensors in the furnace and flue are indicated below:

- Metal temperature sensor - 80 mm above hearth level.
- Flue gas temperature sensor in flue - 0.3 m downstream of the flue offtake centreline.
- Roof refractory temperature sensor - embedded in roof refractory 25 mm behind hot face.
- Furnace pressure sensor - hole in sidewall 0.5 m above metal level.
- O_2 analyser sensor - position as for flue gas temperature.

(b) Instrumentation for Stage I

A common instrumentation package is applied to all four furnaces, but most of the components are purchased from two different manufacturers (Kent Process Control Ltd and Leeds and Northrup Ltd) so that their relative merits can be compared. Briefly, the purpose of each package is to provide continuous monitoring of five basic parameters:

- temperature of metal, flue gas and roof refractory
- furnace pressure
- oxygen content of the flue gas

(c) Control equipment for Stage II

For the development of single-loop automated control of individual parameters in Stage II the sensors and all indicator/controllers purchased for Stage I were used, as these had the required control outputs (4-20 mA). In addition, valve positioners, amplifiers and modulating motors were installed to control furnace pressure, total air/fuel flow and air/fuel ratio.

Stage II single-loop automated control of individual parameters consists of three main loops and some auxiliary safety devices (Fig. 1). Only the functions and principles pertaining to furnace operating procedures and commissioning details are discussed below:

(i) Total air/fuel control

The metal temperature signal is fed to a controller which via the air/fuel valve actuator increases the heat input when the temperature falls below the set-point (Fig. 2a).

(ii) Furnace pressure control

The differential pressure transducer provides a signal to a controller which via the damper air valve actuator controls the air injected into the flue which in turn affects the furnace pressure (Fig. 2b).

(iii) Air/fuel ratio control

The flue gas oxygen analyser provides a signal to a controller which, via a trim valve actuator, adjusts the flow of combustion air to the burners and, thereby, the air/fuel ratio and oxygen content of the flue gas (Fig. 2c).

(iv) Safety devices

Safety devices have been incorporated into Stage II equipment to protect the fabric of the furnace and flue, to minimise heat wastage during furnace operations and to increase the safety of personnel operating the furnaces.

- roof refractory temperature - a thermocouple embedded in the roof brickwork supplies a signal to a controller which causes the furnace burners to be shut down when a safe temperature limit, currently set at 1000°C, is exceeded.
- flue gas temperature - in a similar manner to roof refractory temperature a flue gas thermocouple supplies a signal to a controller which turns the burners to 'low fire' when a safe flue gas temperature is exceeded, (currently set at 1100°C).
- door microswitches - microswitches are provided on door mechanisms and are linked to control actuators, resulting in the burners and air damper operating on their minimum flowrates when the doors are opened, e.g. for charging, making additions, skimming, sampling etc. This control minimises fuel wastage when doors are opened.

Post-installation monitoring includes assessment of the instruments, control systems and burner and furnace operations. Fuel consumptions and savings are evaluated following the implementation of the various stages.

1.3. Results and Discussion

Data collection for Stages I and II on all of the four furnaces, on which some or all single-loop automated controls of independent variables are functioning, is being carried out. Results for fuel savings, melt rates and efficiencies and evaluation of the performance of the instruments and the control equipment are reported below.

During the preliminary surveys on the four furnaces it became clear that all burner systems had the following deficiencies:

(a) They were set and operated incorrectly.

(b) The fuel and air supply control systems were inadequate for ensuring delivery at the specified pressure and flowrate.

(c) In the case of oil fired burners, the oil was not delivered at the correct pressure or temperature, as specified by the burner manufacturer.

(d) Very little of the required routine maintenance was practiced on the oil-fired burners.

(e) The actual turn-down ratio achieved during operation was extremely low.

1.3.1 Chalfont Furnace (experimental)

This furnace has been operating with all single-loop controls since August, 1980. Due to the nature of the operation of this furnace, automated single-loop control of total air/fuel flow had to be incorporated at the commencement of Stage I work. Consequently, the assessment of instrumented manual control includes the use of the above Stage II control feature. Also, some of the fuel consumptions have been compared with the base line data obtained after carrying out the preliminary survey. However, because of the limitations of the previous burners and control systems, fuel savings in this report are mainly based on new base line data obtained on instrumented manual control following the furnace and burner modifications.

1.3.1.1 Fuel consumption during holding period

(a) Nearly empty furnace, containing about 0.25 tonne of molten metal at 725°C (Table I)

(i) Savings based on consumption using previous burners (prior to furnace and burner modifications)

67% .... total savings achieved,

40% .... resulting from new burners and single-loop air/fuel flow control,

27% .... resulting from furnace pressure control.

(ii) Savings based on consumption after furnace and burner modifications (i.e. new burners, instrumentation and automated total air/fuel flow control) (Table 1)

53% .... total savings achieved

45% .... resulting from automated furnace pressure control,

8% .... resulting from reductions in fuel flow settings.

(b) Furnace containing 5 tonnes of molten metal at 720°C (Table II)

Savings based on consumption after furnace and burner modifications (as stated above)

43% .... total savings,

39% .... resulting from automated furnace pressure,

4% .... resulting from reducing the fuel flow settings.

1.3.1.2 Effect of furnace pressure control

(i) By the use of automated pressure control the fuel consumption was reduced from about 1.08 to 0.58 GJ/h, representing a saving of 45% for holding a nearly empty furnace, and from 1.09 to 0.66 GJ/h, representing a saving of 39% for holding a furnace containing 5 tonne of molten metal (Tables I and II).

(ii) The fuel usage increased when the furnace was maintained at a high positive or negative pressure.

(iii) With furnace pressure control the burner thermal inputs and the flue gas temperatures are lower and the magnitude of thermal cycling is significantly reduced.

Based on the above findings and for operating convenience the furnace is normally controlled at a slightly negative pressure (about-0.012 mbar to - 0.025 mbar) and the increase in fuel consumption is not considered significant.

1.3.1.3 Effect of the number of burners

(i) Lower fuel consumptions were achieved when holding a nearly empty furnace by using all the three burners compared with only two e.g. 0.58 compared with 0.90 GJ/h, at zero pressure.

(ii) The effect of using all the burners is less marked when the furnace pressure is not controlled. This is due to the fuel wasted in heating the substantial amount of cold air drawn into the furnace.

(iii) The effect of using 2 instead of 3 burners resulted in higher thermal inputs and flue gas temperature for holding 5 tonnes of metal at 720°C.

From the above results it was concluded that the furnace is best operated with all the burners lit and the benefits are ascribed to:

(i) better distribution of the heat input to the furnace;

(ii) fuel wastage by heating a bleed of air passed through the unlit burners is also eliminated. This bleed was recommended by the manufacturers to avoid cracking of the burner heads by the radiant heat of the furnace.

The burners originally installed in the furnace were achieving a melt rate of 0.8 te/h. The provision of new burners, in conjunction with the fuel savings obtained by the various stages of automated single-loop control, has enabled the fuel flow settings to be reduced without lowering the melt rate. The new burner fuel flow settings on high and low fire are 4.38 and 0.32 GJ/h respectively compared with 6.0 and 1.3 GJ/h for the previous burners.

In general, for furnaces required to hold metal for substantial periods, it is essential that the low fire heat input rate of the burners is below the holding fuel consumption rate to prevent fuel wastage. This is the case with the Chalfont furnace and the corresponding values are 0.32 and 0.5 GJ/h respectively.

1.3.1.4 Melt rates and efficiency

For the experimental furnace at Chalfont the aim has been to maintain the melt rate (i.e. 0.8 tonne/h) achieved with the previous burner system, whilst reducing the fuel consumption. With the new burners and all Stage II control loops in operation, the fuel consumption has been reduced from 5.10 to 4.33 GJ/h tonne and the furnace efficiency increased from 21 to 28%.

The Chalfont furnace control system is being currently modified to use a microprocessor.

1.3.2 Latchford No. 3 melter

Two out of the three control loops are operational. The wall temperature readings, which are used to control total air/fuel flows, are not considered to be satisfactory. This, therefore, involves manual control of the burner throughput on occasions and the need for manual adjustments has prevented the commissioning of air/fuel ratio control. Work is underway on developing a metal temperature monitoring system since the standard thermocouple method is difficult to implement on this furnace.

1.3.2.1 Melt rates and efficiency

(a) Effect of high fire fuel flow setting on melt rate

The optimum high fire oil setting has been found to be 340 litre/h considering the present production requirements, melt rate of about 3.5 tonne/h, and energy usage. The flue gas temperature was lower with this setting and the thermal stresses on the refractories are considered to be less severe compared with the 500 litre/h setting.

The melting fuel consumption has been reduced from 5.8 to 4.1 GJ/tonne. Fuel saving achieved is 29% divided as follows:

- due to instrumented manual control - 13.5%
- due to automated furnace pressure control - 15.5%.

Savings evaluated on individual melts have been as high as 50%.

Improved burner settings and instrumented manual control have increased the furnace efficiency from 22.7 to 33%.

(b) Melt rates

The actual melt rates ranged from 2.4 to 3.7 tonne/h, with delay periods up to 3¼h. The calculated melt rate, excluding delay periods, ranged from 3.0 to 4.6 tonne/h. The melting furnace is operated in the holding mode during these delay periods and, as the burners do not have a high turn-down, this results in a higher fuel consumption than would normally be expected.

Improvements to fuel and air supplies to the burners, and the automatic control of the air damper system, have brought the control of oxygen content in the flue gas, on high and low fire (i.e. 1.5 and 4.5% respectively) within acceptable limits. The current turn-down is about 3.5:1.

1.3.3 Latchford No 3 holder

All the three control loops are operational. Depending on production requirements, this furnace is also used as a melter.

1.3.3.1 Fuel consumption for holding

(i) The overall fuel consumption varies between 100 to 180 GJ/day depending upon whether the furnace is used for holding operations only or if a significant quantity of metal is added and melted in the holder, as is the current production practice.

(ii) Fuel savings (consumptions) achieved for holding operation only

- after improved manual control - 21%(revised consumption 79 GJ/day)
- after instrumented manual control - 45%(revised consumption 55 GJ/day).

A large amount of air enters the furnace due to inadequate sealing around furnace doors, which are in a poor state of repair. Consequently, further savings are anticipated once the furnace doors are repaired.

1.3.4 Falkirk No 11 holder

(a) Current status of furnace

Commissioning of Stage II (automated single-loop control) equipment is complete and the furnace now operates under automatic control of the following factors:

(i) Total air/fuel flow
(ii) Air/fuel ratio
(iii) Furnace pressure (by air curtain damper mounted in flue).

(b) Fuel consumption

The observed daily fuel consumption under normal production conditions varied between 21.1 and 22.2 GJ/day, a reduction of about 33% when compared with the data collected before implementation of the programme, i.e. about 32.7 GJ/day.

(c) Heat input rate

Previous studies of the furnace indicated that a reduction in the heat input at high fire could be made without any adverse effect on furnace operation. The high fire heat input was, therefore, gradually reduced from 3.7 GJ/h to 2.8 GJ/h. The turn-down achievable with the system is now 13:1 compared with the 4:1 observed during preliminary survey work. This has been possible by improved burner settings and the provision of air/fuel ratio control. The low fire heat input has, therefore, been reduced from 0.95 GJ/h to 0.21 GJ/h.

(d) Combustion conditions

On high fire, the burners are set to operate at the optimum condition of 1% oxygen, in the flue gas, i.e. 5% excess air. On low fire the oxygen content is quite high, i.e. 11%, this being due to the design of the burners.

1.3.5 Estimated fuel savings

(a) Chalfont furnace

As this furnace is an experimental unit in the BA Research Laboratory and is not used continuously or repetitively under the same conditions, energy savings have not been computed.

(b) Latchford No 3 melter

The fuel consumption has been reduced from 5.8 to about 3.2 GJ/tonne, due to instrumented manual control and automatic pressure control. The saving achieved due to the above is about £5.00 per tonne and on a melt throughput of about 8,000 tonne per annum, the total saving is estimated to be about £40,000 per annum.

(c) Latchford No 3 holder

This furnace is generally used as a conventional holder, but there are periods when melting is also carried out. Fuel savings are based on this furnace being used as a conventional holder only. Fuel consumptions have been reduced from 100 to 55 GJ/day, a saving of 45 GJ/day which is equivalent to £110 per day and an annual saving of about £25,000 for a 45 week working period.

(d) Falkirk No 11 holder

Fuel consumption on this furnace has been reduced from 32.7 to 21.5 GJ/day, a saving of about 11 GJ/day which is equivalent to £20 per day and an annual saving of about £7,000, based on a 45 week working period.

1.3.6 Assessment of Instrumentation and Control Equipment Performance

Some of the assessments made of the instrumentation and control equipment and the burners are given below:

1.3.6.1 Total air/fuel flow control

(a) Sensors - No major problem is encountered with the sensors used, i.e. chromel/alumel thermocouples used for metal temperature measurements, but short lives are obtained with the refractory protective sheaths used, due to mechanical damage, particularly during charging in the melters.

(b) Controllers

The Control and Readout Ltd (CRL) Model 405 unit has all the desirable operational features and is recommended despite its slightly higher cost. Other units are satisfactory, particularly Eurotherm Model 070. Some difficulties were experienced with Leeds & Northrup (L & N) Electromax units and it is concluded that these are not ideally suited to this application.

(c) Problems encountered

The amount of the temperature overshoot above the control set point has been a major problem.

There are two parameters to be considered - the magnitude and duration of the temperature overshoot both of which increase with the bath depth. Careful selection of the PID (Proportional, Integral and Derivative) functions of the controller are essential to prevent or minimise the magnitude and duration of the overshoot.

The CRL 405 controller is recommended and the Eurotherm 070 controller is satisfactory for the application.

(d) Control mode

During charging the thermocouple is usually removed from the furnace, to prevent mechanical damage to its sheathing. During this period the temperature reading is well below the set point and the air/fuel flow to the burners is at a maximum. Control of the air/fuel flow during melting and holding is therefore not possible until the thermocouple, with its sheathing, is re-introduced to the furnace containing a heel of molten metal. The major disadvantage of this method (Fig. 2a) is the lack of control during periods when the thermocouple is out of the furnace. Flue gas temperature is now used to control total air/fuel flow to the burners during melting operations when the metal thermocouple is normally withdrawn (Fig. 2e). After the thermocouple is re-inserted and temperature rises to a pre-set level, transfer to metal temperature control occurs automatically when a second set point using metal temperature is activated. This system allows a control loop for total air/fuel flow to operate during the whole of the melting cycle even when the metal temperature sensor has to be removed to prevent damage.

1.3.6.2 Furnace pressure control

Few problems have been encountered provided the system is operated with care. The following factors have to be considered in order to maintain the life and efficiency of the system.

(a) damage to damper manifolds - if the damper air supply is not maintained during furnace operation.

(b) damping - a restricting valve or orifice in the sensor probe may be necessary to damp down fast oscillations in furnace pressure caused by burner characteristics.

(c) instrumentation - pressure transmitters must be robust, responsive and accurate. Kent and L & N units have been successfully used and are recommended but some spares availability problems have been encountered with the latter.

The Foster Cambridge and L & N controllers do not have the required PID functions. The new CRL 407 has this function and this controller has been used in a similar application.

(d) location of the probe in the furnace.

1.3.6.3. Air/fuel ratio control

Three oxygen analysers have been evaluated from Kent, L & N and Neotronics. All three operated satisfactorily but the convenience, size, availability and low price of the Neotronics make it highly recommended.

(a) Oxygen analysis equipment

Although all the types of analyser tested have performed well, much effort has had to be spent in order to develop sampling systems which are reliable, effective and suited to the degree of contamination encountered in the flue gases. The flue gases in oil fired furnaces comprise CO_2, O_2, H_2O, N_2, SO_2 and chlorides of sodium and potassium. In addition, in the melters where lacquered scrap is melted HCl or HF are evolved. In gas fired melters and holders, SO_2 is not present in flue gases. In addition to the toxicity of the above constituents large quantities of alumina dust and soot are present. The conditions in the flues in the melters are more demanding on the stainless steel sampling probe and the oxygen sampling system compared with the holders. The major problems encountered so far are corrosion and blockage of the probe and ineffective scrubbing. For Chalfont and Falkirk furnaces (gas-fired) a system was developed which operated successfully. For No. 3 oil-fired melter at Latchford, however, a commercial sampling and scrubbing system was purchased from 'L & N'but this system suffered a series of problems indicated above. The system was modified into a Chalfont developed system which is operating satisfactorily.

1.3.6.4 (b) Other instruments and equipment

The various types of valve positioners and control motors used in the control equipment have operated satisfactorily.

1.3.6.5 Air damper system

The system consists of an air fan, a motorised valve and a manifold located around the perimeter of the flue section. Two types of design have been used in which motorised valve is located either before or after the manifold. Because little damper effect is required to maintain a positive furnace pressure under 'high fire' conditions, the first type of design requires that a small bleed of air be passed through the manifold at all times to avoid it becoming damaged by overheating. However, with the second design type, full air flow is maintained through the manifold at all times, since flow is vented at high fire; therefore, this type of damper is more suited to the stringent requirements of factory melting furnaces (e.g. Latchford No. 3) while the first arrangement can normally be successfully applied on holders. The air dampers have performed very well both in vertical and in below ground horizontal flues.

1.3.6.6 Burners used in Chalfont furnace

The two Nu-Way Multiplex MP4 burners and one MP2 burner have operated well. They have a very high turn-down ratio (14:1) based on fuel flow and 'on-ratio' (1.3% O_2). MP4 burners were developed for this application by Nu-Way Energy Ltd (previously Nu-Way Eclipse Ltd) as they did not have a burner for the range required. The previously available MP2 burners were of smaller capacity. Initial developments were carred out at Chalfont and now Nu-Way have MP3 and MP4 burners available for a range of capacities. Considerable difficulties were experienced during the initial development period in the following areas:

- flame instability on low fire.
- achieving turn-down ratios greater than 10:1 based on fuel flow.
- achieving and maintaining the desired combustion conditions.

1.3.7 Stage III - Fully automated multi-loop control

Equipment requirements for Stage III fully automated control have been assessed and the components for a microprocessor based control system purchased. An evaluation study was carried out in which factors like capital and revenue costs and time for development and implementation were considered. It was decided to opt for a small computer to provide management information and an independent control unit dedicated to a furnace in particular. Other options which were discounted were to have either an integrated unit capable of handling both the requirements or one control microcomputer to control multiple furnaces.

Two components were considered during the development stage:

(a) A small computer programmed for all mathematical operations and for providing management information.

This was given the <u>first</u> priority and the small computer was used during Stages I and II in BA factories for calculations and provision of the necessary information.

(b) A system for automatic combustion control.

Three phases of development are envisaged for microprocessor control:

<u>Phase 1</u> - a unit to by-pass existing equipment installed on the Chalfont furnace.

<u>Phase II</u> - provision of a similar unit for installation at Latchford.

<u>Phase III</u> - specification of a "black-box", microprocessor control unit for combustion control.

Phase I, on the Chalfont furnace, is underway. For Phase II the microprocessor unit is being constructed and the necessary modifications to the controller is in progress. The existing control equipment has required further wiring and when this is finished the microprocessor will be installed.

The equipment selected for the two components are given below:

(i) <u>for the provision of management information</u>

An Apple II Microprocessor package has been purchased. The specification included twin disc drive units and a line printer. The computer is equipped to operate with either U.C.S.D. PASCAL or BASIC languages and incorporates a 48K memory.

(ii) <u>for automatic combustion control system</u>

The system chosen as the basis for this unit was the Intel 8085 Microprocessor and standard modules have been used. The system will have the flexibility to incorporate the furnace control strategy following the initial period of monitoring and displaying of primary furnace parameters.

<u>CONCLUSIONS</u>

The findings to date on the work carried out on Stages I and II are summaries below:

(a) <u>Fuel savings</u>

It has been shown that large fuel savings can be achieved by the application of the following measures:

- improved burner specifications, particularly increased turn-down ratio,
- application of instruments,
- automated single-loop control of individual parameters.

A combination of some or all of these measures has given the following fuel savings:

(i) <u>Chalfont furnace</u>

The savings achieved during holding, taking the original burner system, prior to any modifications, as the datum line, were:

67% .... total savings

40% .... due to the improved burners

27% .... due to pressure control

Taking the new datum line of improved burners, instrumentation and automated total air/fuel flow control redefines the savings on holding as:

53% .... total savings

45% .... due to automated pressure control

8% .... due to reduced fuel flow

For melting, using the datum line of furnace prior to any modifications savings up to 14.5% have been obtained.

(ii) Production furnaces

45% .... for Latchford No 3 holding furnace with instrumented manual control

50% .... savings have been recorded for individual melts

30% .... average savings during normal operation with instrumented manual control and automated pressure control

A summary of the current level of fuel savings is as follows:

- Latchford No. 3 melter - Reduction in fuel usage from 5.8 to 3.2 GJ/tonne. Annual fuel saving estimated £40,000.
- Latchford No. 3 holder - Reduction in fuel usage from 100 to 55 GJ/day. Annual fuel saving estimated £25,000.
- Falkirk No. 11 holder - Reduction in fuel usage from 32.7 to 21.5 GJ/day. Annual fuel saving estimated £7,000.

(b) Optimisation of operating conditions

The work carried out has highlighted the following aspects of furnace operation relating to the efficient use of fuel.

(i) The Chalfont furnace may be operated most efficiently at a slightly negative pressure (- 0.012 mbar to - 0.025 mbar). Other furnaces may require a different furnace pressure setting because of design differences.

(ii) Furnaces may be operated most efficiently with all the available burners lit.

(iii) A burner system with a high turn-down ratio is considered to be a major factor in achieving fuel savings. This was particularly apparent on the furnaces in which both melting and holding were carried out, because of the widely differing heat input requirements of the two modes of operation.

(c) Instrumentation and control systems

Most of the control loops are now fully operational, and modification of melter air/fuel flow controls are proved to be of benefit.

Assessments carried out on the equipment and instrumentation have shown that modification of certain components has been necessary in order to provide satisfactory performance. Particular areas of attention have been oxygen sampling systems and control instrumentation 'PID' functions.

The evaluation of the instrumentation and equipment has clearly shown that some of the equipment commercially aimed at the furnace control market is better suited to the particular rdquirements of this project work than others. Of the numerous indicator/controllers tested, those of CRL proved particularly successful in operation and the Neotronics OTOX 90 oxygen analyser is also worthy of particular note.

DISCUSSION

Question - Chairman : What is the anticipated payback period ?

Answer 6-8 Months. The cost of the modifications is of the order of £15000/unit (27000 ECU/unit)

Question - Chairman : Is there scope for self-recuperative burners on these furnaces ?

Answer Yes, we are working in conjunction with British Gas on this.

Question : Did not the manufacturer of the furnace provide a control system ?

Answer Yes, but we have improved on the manufacturer's specification. Later, we may convince manufacturers to use our system.

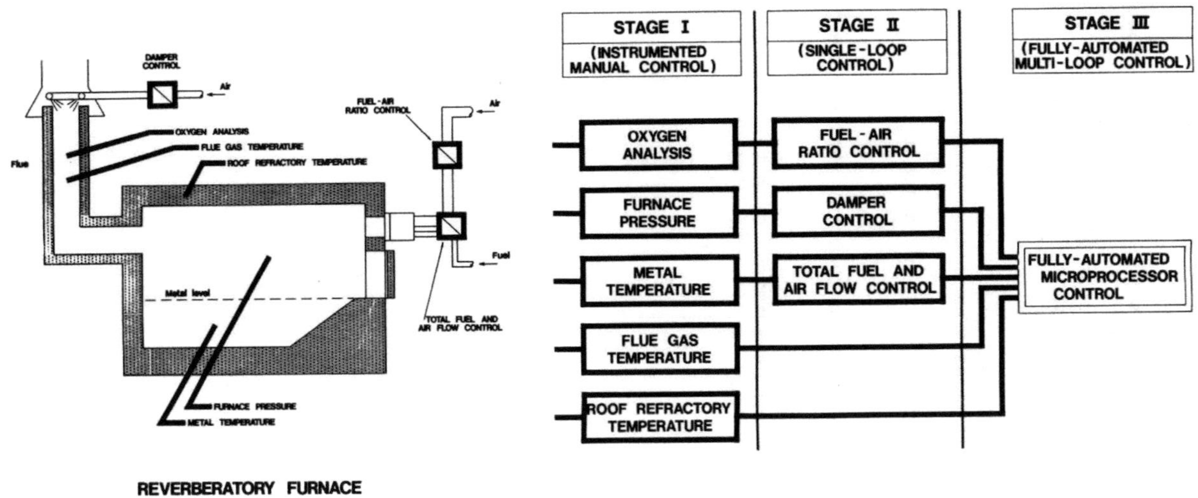

Fig. 1 Automated combustion control system (schematic)

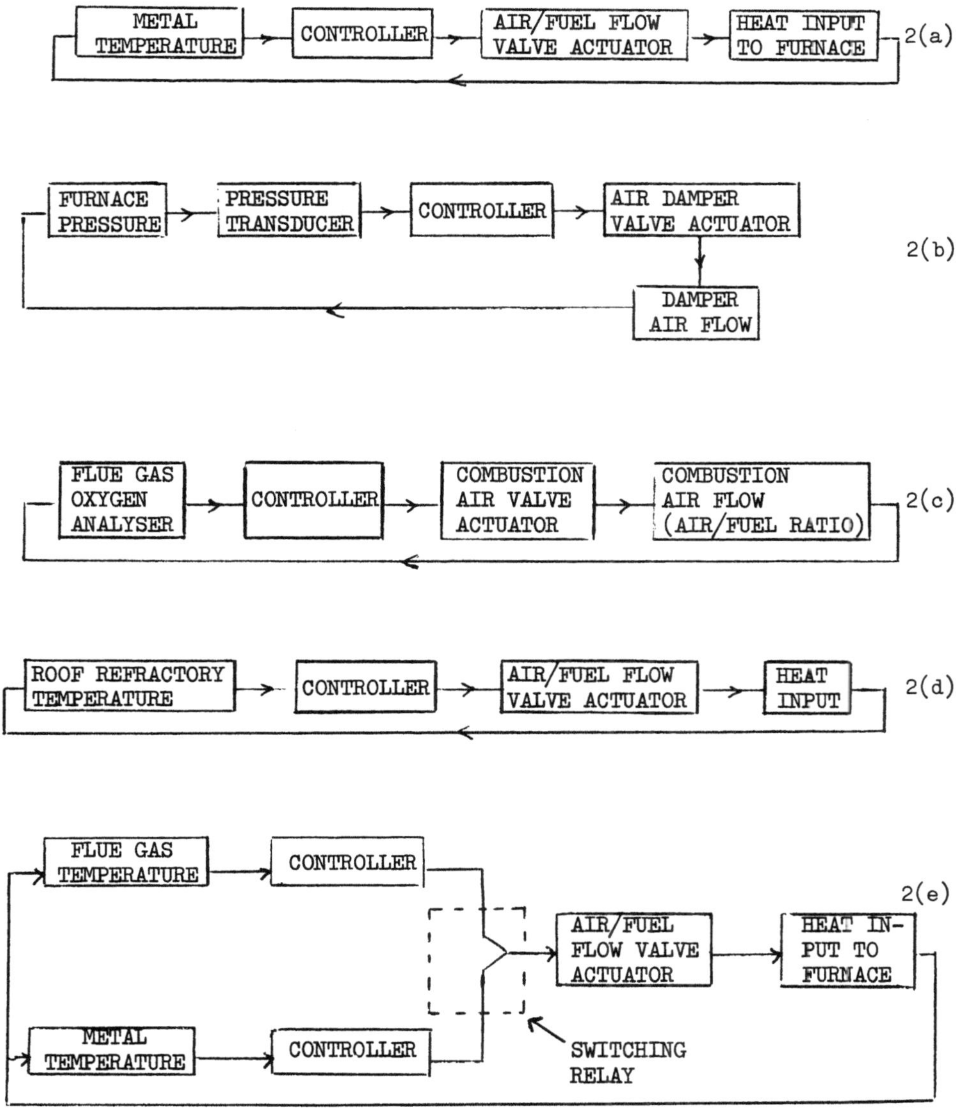

2(a) Total air/fuel flow control using metal temperature
2(b) Furnace pressure control
2(c) Air/fuel ratio control
2(d) Roof refractory temperature control
2(e) Total air/fuel flow control using flue gas/metal temperatures

Fig. 2 Single-loop control systems

TABLE 1

Fuel consumptions for holding a nearly empty furnace (0.25 tonne of molten metal) at 720°C

(Chalfont 8 tonne furnace)

Description	Furnace pressure (mbar)	Fuel consumption (GJ/h)	Fuel savings (%) based on	
			(1)	(2)
Previous burners (prior to furnace and burner modifications$^{(1)}$)	- 0.17	1.79	-	-
Instrumented namual control after furnace and burner modifications$^{(2)}$ (i.e. new burners, instrumentation and total fuel/air flow automatic control)	- 0.17	1.08	40	-
After automated furnace pressure control	-	0.58	67	45
After reductions to the fuel flow settings	-	0.5	-	53

TABLE II

Fuel consumption for holding 5 tonne of molten metal at 720°C

(Chalfont 8 tonne furnace)

Description	Furnace pressure (mbar)	Fuel consumption (GJ/h)	Fuel savings based on (1) (%)
Instrumented manual control after furnace and burner modifications (i.e. new burners, instrumentation and total fuel/air flow automatic control)$^{(1)}$	- 0.17	1.09	-
After automated furnace pressure control	-	0.66	39
After reductions to the fuel flow settings	-	0.62	43

IMPROVEMENTS OF RECUPERATORS ASSOCIATED TO STEEL SLAB FURNACES

Author	: J.L. ROTH
Contract number	: EEB-1-150-F
Duration	: 24 months 1. January 1981 - 31 December 1982
Head of Project	: J.L. ROTH, Chef du Service "Fours", IRSID.
Contractor	: IRSID - Institut de Recherches de la Sidérurgie Française
Address	: 57210 MAIZIERES-lès-METZ - France

SUMMARY

The reheating furnaces before the mills for steel strips and plates represent important fuel consumers : a furnace of this type has a yearly consumption of 40 000 to 50 000 tons of equivalent-petrol.

The aim of this study consists of decreasing this consumption, by improving the efficiency of the recuperators which receive the waste gas leaving the furnace, and preheat the combustion air.

These improvements concern mainly the radiation recuperators ; the most french slab reheating furnaces are equiped by this type of recuperator, with a poor efficiency. Furthemore the radiation recuperators will remain the first part of the new recuperation installations, because of the high waste gas temperatures at the outlet of the furnaces.
The work done consists of the following steps :

- construction and experimentation of a recuperator model (scale 1:5) ;
- elaboration of a calculation programm of a recuperator, and its adjustment on the base of practical measurements ;
- measurements on a work recuperator ;
- methodic thermal analysis of the observed phenomena, leading to suggestions for improvements.

In a next step, these ideas of improvements will be tested on the model.

I. PRESENTATION DES FOURS A BRAMES ET ENJEU ENERGETIQUE DES RECUPERATEURS

Les fours qui équipent les laminoirs à produits plats d'acier (trains à bandes ou à tôles fortes) sont des outils de masse et de gros consommateurs d'énergie : ainsi, un train à bandes produisant 4 millions de tonnes d'acier par an comporte en général 3 à 4 de ces fours, dont chacun consomme environ 40 000 à 50 000 tep. (tonne - équivalent de pétrole)de gaz ou de fuel lourd par an.

La planche (1) présente un schéma d'ensemble d'un four de ce type, ainsi que les éléments principaux de son bilan thermique.

Les "brames" d'acier y sont réchauffées à des températures élevées (1200°C) et les fumées sortant de l'enceinte de combustion, sont elles-mêmes à des températures de l'ordre de 1000°C. Une partie du potentiel thermique de ces fumées est cependant recyclée par l'intermédiaire d'un échangeur placé sur le conduit des fumées, servant à préchauffer l'air de combustion envoyé aux brûleurs : le principe de ce "récupérateur" à air chaud est également présenté à la planche (1) qui le situe dans le contexte énergétique d'ensemble d'un four à brames.

On retiendra que les améliorations des récupérateurs actuels peuvent induire des économies de combustible de l'ordre de 1000 à 5000 tep. par four et par an.

2. LES GRANDS TYPES DE RECUPERATEURS ET LEURS PERFORMANCES ACTUELLES ; INTERET DES RECHERCHES DANS CE DOMAINE

Les technologies des récupérateurs utilisés actuellement peuvent être réparties en 3 familles ; la planche (2) en présente des schémas de principe :

- les récupérateurs céramiques, contitués d'empilages réfractaires, admettent des fumées atteignant 1400°C, mais sont handicapés dans leurs performances et dans leur exploitation par la contrainte du fonctionnement en dépression (ventilateur d'air situé entre le récupérateur et le four). Ces récupérateurs sont en voie d'être abandonnés ;

- les récupérateurs métalliques à rayonnement qui équipent la plupart des fours à brames français sont constitués d'une grosse conduite d'évacuation des fumées et d'une conduite concentrique externe, l'air parcourant l'espace annulaire ; ces appareils admettent des fumées à des températures allant jusqu'à 1100°C, mais leurs rendements actuels sont inférieurs à 30 %, malgré leurs dimensions imposantes (hauteurs de l'ordre de 30 m) ;

- les récupérateurs tubulaires à convection sont composés de tubes métalliques minces traversant le conduit des fumées et alimentés par de l'air à grande vitesse ; la tenue des tubes limite à environ 900°C la température des fumées à l'entrée ; aussi ce type d'appareil est-il parfois mis en série avec un récupérateur à rayonnement ("récupérateur mixte"). Pratiquement, les récupérateurs à convection conduisent toujours à des rendements de récupération satisfaisants (40 %), voire très bons (70 %) lorsqu'on les dote d'une grande surface d'échange.

Pour améliorer la récupération sur les fours actuels, il ne suffit pas pour autant de remplacer leurs récupérateurs à rayonnement par des appareils à convection. En effet, les températures de fumées pratiquées dépassent souvent 1000°C et la protection des tubes convectifs ne peut être assurée de manière satisfaisante par la seule dilution des fumées : ceci conduirait à perdre sur l'efficacité de l'échange ou à surdimensionner l'appareil.

Pratiquement, l'on est souvent amené à appliquer la solution mixte, consistant à protéger le récupérateur à convection par une première partie à rayonnement.

Mais, cette solution conduit à un investissement élevé (de l'ordre de 10 % du prix d'un nouveau four) et à des travaux de grande envergure (immobilisation d'un four sur au moins 1 mois) qui ne pourront être réalisés sur les fours d'une batterie que de manière échelonnée dans le temps (sur une période de 5 ans par exemple).

Il apparaît alors doublement intéressant d'étudier le fonctionnement et les améliorations possibles des appareils à rayonnement actuels :

- pour essayer d'améliorer leurs performances à court terme, avant leur remplacement ;

- pour définir au mieux la partie à rayonnement d'une installation de récupération mixte lors du changement de récupérateur.

Enfin, nous pensons que les leçons de caractère fondamental de nos études sont susceptibles d'induire également des améliorations dans la conception des récupérateurs à convection, à moyen ou à long terme.

3. MAQUETTE DE RECUPERATEUR, EXPERIMENTATION ET MODELISATION; RESULTATS ACQUIS ET PERSPECTIVES D'AMELIORATION DES RECUPERATEURS D'USINE

L'analyse complète des objectifs et méthodes de la recherche nous a conduits à concevoir et à construire à l'IRSID une maquette de récupérateur à rayonnement.

Nous avons développé en même temps un modèle de calcul des échanges de chaleur dans les récupérateurs à rayonnement.

Ce modèle a été confronté successivement à un grand nombre d'essais sur la maquette, ainsi qu'à une série de mesures effectuées sur un récupérateur d'usine.

Enfin, l'analyse méthodique des résultats ainsi obtenus nous a conduits à imaginer des moyens d'améliorer les échanges entre les fumées et la paroi refroidie par l'air ; l'étroitesse de l'espace annulaire parcouru par l'air laisse peu d'initiative à une amélioration des échanges du côté de l'air.

3.1. Présentation de la maquette de récupérateur :

La maquette de récupérateur à rayonnement construite à l'IRSID est présentée à la planche (3) :

- il s'agit d'un récupérateur à rayonnement pur, constitué de deux tubes concentriques, les fumées parcourant le tube interne, l'air étant réchauffé dans l'espace annulaire.

- la première partie de l'échangeur est alimentée en air à co-courant, la deuxième à contre-courant, comme c'est le cas pour les récupérateurs industriels.

- la figure 1 présente le schéma de l'installation, ainsi que ses dimensions caractéristiques :

* hauteur : 3 m (1/5 à 1/10 de la hauteur des appareils industriels),
* diamètre extérieur : 0,80 m,
* espace annulaire : 6 mm.

- la maquette ne respecte pas une échelle géométrique par rapport à un récupérateur industriel : ses dimensions ont été définies pour conserver les niveaux d'échanges thermiques dans chacune des parties :

* conservation du temps de séjour moyen des fumées dans le tube interne (échange par rayonnement largement prépondérant) ;
* conservation des vitesses de l'air dans l'espace annulaire (échange par convection).

3.2. Expérimentation de la maquette et comparaison avec le modèle :

Nous avons procédé sur la maquette à un grand nombre d'essais qui nous ont permis successivement :

a) de mettre au point et de rationaliser les techniques de mesures pour les essais ultérieurs sur les installations d'usine.

Des essais effectués, on peut extraire un point de fonctionnement moyen de la maquette :

* four alimenté en gaz naturel,
* débit de fumées 1200 nm^3/h,
* débit d'air 1100 nm^3/h,
* température des fumées :
  + à l'entrée du récupérateur à co-courant : 1000°C,
  + à la sortie du récupérateur à contre-courant : 770°C.
* réchauffage de l'air de 40°C à 280°C (370 nm^3/h dans la partie supérieure à contre-courant, 730 nm^3/h dans la partie inférieure à co-courant).

b) d'ajuster le modèle de récupérateur à rayonnement que nous avons élaboré :

* le modèle utilisé est un modèle "par tranches", utilisant les formules classiques du transfert de chaleur par rayonnement et par convection, et procédant par itération pour le calcul de l'équilibre thermique qui s'établit dans chaque tranche entre les gaz et les deux parois ;

* après prise en compte du gradient de température mesuré dans l'épaisseur du courant des fumées (et sur lequel nous reviendrons) et des pertes thermiques par la paroi externe (10 à 15 % de l'échange global dans le cas de notre maquette), nous avons observé une bonne concordance entre le calcul et la mesure, l'écart constaté étant en général inférieur à 5 % (10 à 15°C sur la température de l'air).

c) d'analyser certains problèmes pratiques de dissymétrie de température apparus sur la circonférence de la maquette et qui peuvent également fausser les mesures et interprétations sur les appareils d'usine.

d) de quantifier l'incidence sur le rendement du récupérateur des problèmes pratiques d'exploitation, tels que les pertes de fumées, les entrées d'air, ou un excès d'air trop élevé aux brûleurs.

e) de "découvrir" par des mesures fines un phénomène aérothermique inattendu : dès le démarrage des essais, nous avons observé un fort gradient de température dans l'épaisseur du conduit des fumées, à l'entrée et à la sortie du récupérateur à contre-courant ; l'allure et l'importance des gradients de température ainsi mesurés sont présentées à la planche (4).

Ces gradients doivent être évidemment pris en compte pour la détermination d'une température moyenne des fumées dans une section du conduit ; mais il convenait en outre de les expliquer, et de les retrouver sur un appareil industriel.

3.3. Mesures sur un récupérateur d'usine :

Dans un deuxième temps, des mesures ont été effectuées sur un récupérateur industriel équipant l'un des fours de SOLMER à Fos-sur-Mer.

Un schéma de ce récupérateur est présenté à la planche (5).

Bien que moins complètes que celles effectuées sur notre maquette, ces mesures ont permis néanmoins de vérifier la bonne représentativité de la maquette, tant au plan des transferts globaux de chaleur que du phénomène aérothermique décrit précédemment.

3.4. Analyse méthodique des phénomènes observés :

Il convenait d'analyser avec soin ces résultats et phénomènes observés, pour en déduire des essais d'amélioration pratique du rendement de l'échangeur. C'est ce que nous avons fait, en nous appuyant sur la littérature classique concernant les échanges convectifs et radiatifs.

3.4.1. Analyse des échanges convectifs dans le conduit des fumées

A la découverte du gradient de température mesuré dans l'épaisseur du courant de fumées, nous nous sommes posé deux questions :

a) ce gradient important est-il "normal", c'est-à-dire compatible avec les lois classiques de l'aérothermique ?

b) si oui, peut-on espérer modifier ce gradient, et améliorer ainsi le transfert de chaleur ?

Nous avons pu répondre à la question **a)** en retraçant un profil thermique complet de la température des fumées et de la paroi (adjacente en se plaçant par exemple à mi-hauteur de notre maquette, on est en présence de fumées à 850°C au centre de la conduite, la paroi étant à environ 500°C),et en appliquant à ce profil l'analogie de Martinelli entre vitesses et températures pour un écoulement turbulent établi, en tube cylindrique, de gaz dont le nombre de Prandtl est voisin de 1 : dans ce cas, Martinelli montre en effet qu'on peut superposer le profil des températures au profil des vitesses. Ceci est présenté à la planche (6).

Pour aborder le problème **b)**, il faut se pencher sur la mise en régime de l'écoulement gazeux à l'entrée du tube : cette mise en régime est schématisée qualitativement à la planche (7). Quantitativement, Mc Adams prévoit une longueur d'établissement du régime turbulent donnée par des formules du type :

$$\frac{L_{\text{établie}}}{D} = f\ (Re)$$

où la fonction f dépend de la géométrie de l'entrée du tube (entrée en forme de cloche, avec ou sans coude, etc.).

En choisissant une formule associée à une géométrie de l'entrée proche de celle d'un récupérateur, on trouve des longueurs d'établissement représentant plusieurs fois le diamètre de la conduite des fumées, soit une part notable de la hauteur du récupérateur.

Par ailleurs, il en résulte, dans la zone non établie, un <u>accroissement du coefficient d'échange convectif</u> d'autant plus important que le nombre de Reynolds est élevé :

$$\frac{h}{h_{\text{établi}}} = 1 + \frac{0{,}15\ Re^{0,25}}{L/D}$$

L'application de cette formule - très empirique - donne :

- dans le cas de la maquette ($Re \sim 16\ 000$), un accroissement de l'ordre de 30 % de l'échange convectif ;
- dans le cas du récupérateur d'usine ($Re \sim 200\ 000$), un accroissement de près de 60 % de l'échange par convection ; mais rappelons que celui-ci ne représente que 15 % environ de l'échange global dans un récupérateur à rayonnement.

Nous retiendrons que le régime transitoire à l'entrée de la conduite des fumées est bénéfique, et qu'on peut donc essayer de l'entretenir pour améliorer encore l'échange convectif entre les fumées et la paroi (bas de la planche (7) ; nous essaierons plus loin d'imaginer des dispositifs permettant de tester cette idée sur notre maquette.

3.4.2. Examen de l'échange radiatif

L'échange radiatif étant largement prépondérant (plus de 80 %) dans le transfert global de chaleur d'un récupérateur à rayonnement, il est tentant d'essayer de l'accroître également (ne serait-ce que de quelques % !).

Nous avons donc pensé à appliquer au récupérateur à rayonnement une idée concrétisée -avec succès- dans le tunnel d'entrée de certains fours de faible capacité non équipés de récupérateurs : il s'agit de placer dans le courant gazeux des panneaux métalliques ou réfractaires servant de relais au transfert radiatif (et convectif) entre les fumées et la paroi refroidie par l'air. Le principe en est présenté à la planche (8) : on constate qu'un tel panneau -relais sera d'autant plus efficace qu'il est moins refroidi par son ancrage à la conduite (ce qui ira évidemment à l'encontre de sa tenue !).

Voyons à présent s'il existe des moyens technologiques simples de réaliser ces idées dans la pratique.

3.5. Dispositifs pratiques améliorant le rendement des récupérateurs :

Pour tester les idées présentées ci-dessus, nous allons entreprendre (planche (9)) successivement les modifications suivantes de la maquette :

a) installation d'un "cône - turbuleur" devant entretenir le régime d'écoulement transitoire à l'entrée du récupérateur et accroître ainsi le transfert convectif paroi - fumées ;

b) mise en place d'un "croisillon - relais" dans le but d'augmenter l'échange radiatif paroi - fumées ;

c) introduction d'une "spirale" convecto - radiative qui pourrait cumuler les deux effets.

4. OBFECTIFS A MOYEN ET A LONG TERME

Dans la mesure où l'une des modifications que nous nous proposons de tester sur la maquette se révèlerait à la fois efficace et simple à réaliser, nous serions en mesure de la dimensionner pour un récupérateur d'usine.

Cependant, les récupérateurs du type "escher" qui équipent les fours à brames français comportent une partie supérieure munie d'ailettes (planche (5)) et l'analyse des échanges thermiques est à refaire pour cette partie ; nous avons donc engagé une modélisation fine des phénomènes aérothermiques en jeu dans ce cas particulier.

Nous espérons donc être en mesure à moyen terme de proposer aux usines concernées, une série de modifications constructives du récupérateur permettant d'améliorer les échanges du côté des fumées.

A plus long terme, nous pensons examiner également les échanges du côté de l'air, en recherchant notamment des garnissages permettant d'augmenter le coefficient d'échange par convection.

Il faudra cependant tenir compte d'un facteur supplémentaire : du fait de l'étroitesse de l'espace annulaire, la perte de charge du côté de l'air est déjà élevée (400 à 600 mm de colonne d'eau), et le surcroît de perte induit par la mise en place d'obstacles impliquera probablement le changement du ventilateur d'air.

PLANCHE (1)

FOURS DE RECHAUFFAGE

Schéma d'ensemble et enjeu énergétique des récupérateurs

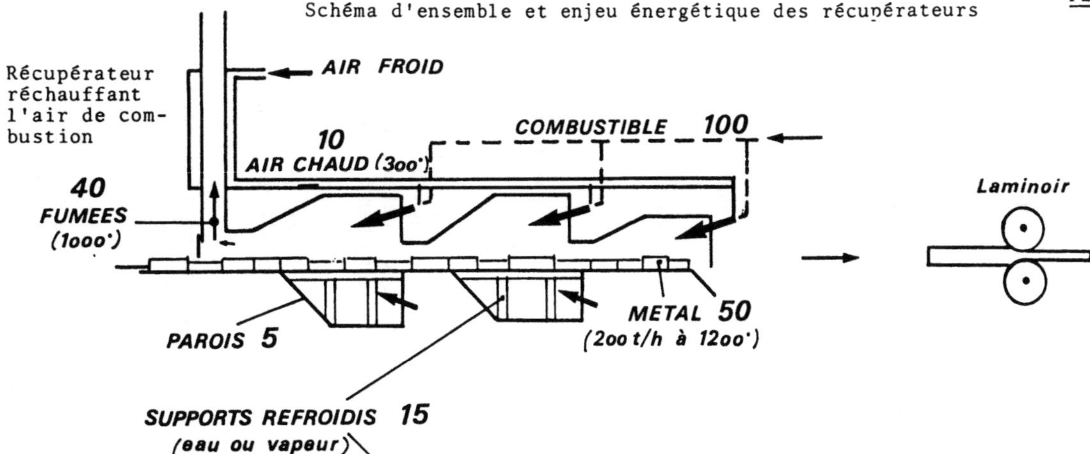

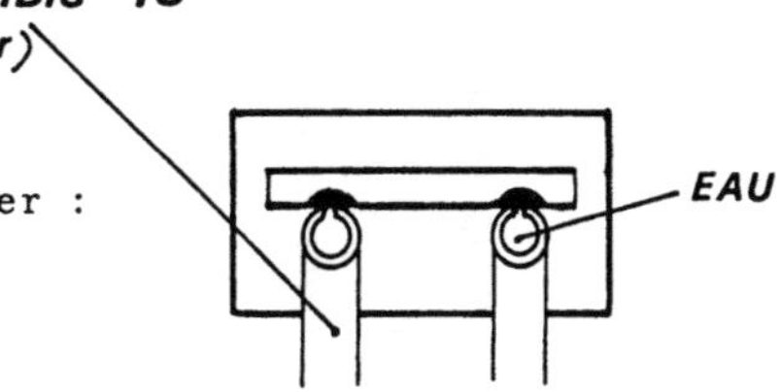

Longueur du four : 35 m ;
largeur : 13 m ;
dimension des "brames" d'acier :
0,2 m x 1 m x 12 m

Production annuelle d'un four : 1 million de tonnes

Consommation actuelle : 400 th/t

soit l'équivalent de 40 000 tonnes de pétrole/an

> Avec un récupérateur réchauffant l'air de combustion à 600°C au lieu de 300°C, on gagnerait 10 % sur la consommation de combustible, soit l'équivalent de 4000 tonnes de pétrole par four et par an.

Coût énergétique en F/t du réchauffage-laminage

combustible (four)

400 th/t x 10 c/th = 40 F/t

électricité (laminoir)

120 kWh/t x 1 5 c/kWh = 18 F/t

(Prix de revient de la tonne de produit fini : 1000 F/t)

LES TROIS TECHNOLOGIES DE RECUPERATEURS

PLANCHE (2)

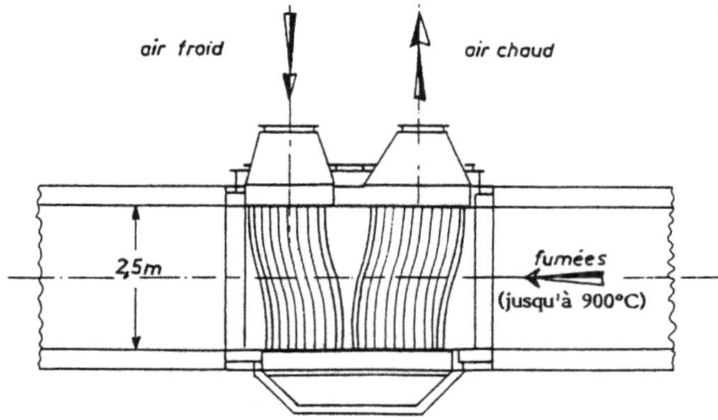

Récupérateur à convection.multitubulaire

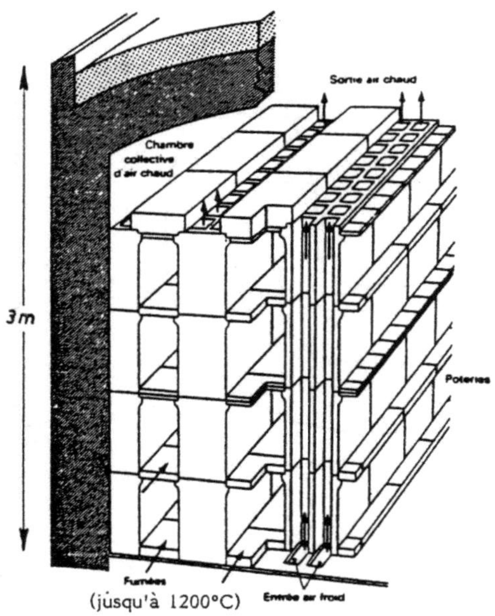

Récupérateur céramique.

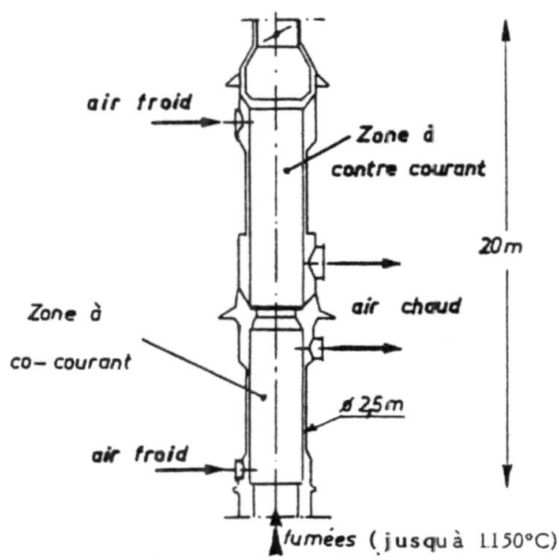

Récupérateur métallique à rayonnement.

PLANCHE (3)

MAQUETTE IRSID DE RECUPERATEUR

-schéma de principe-

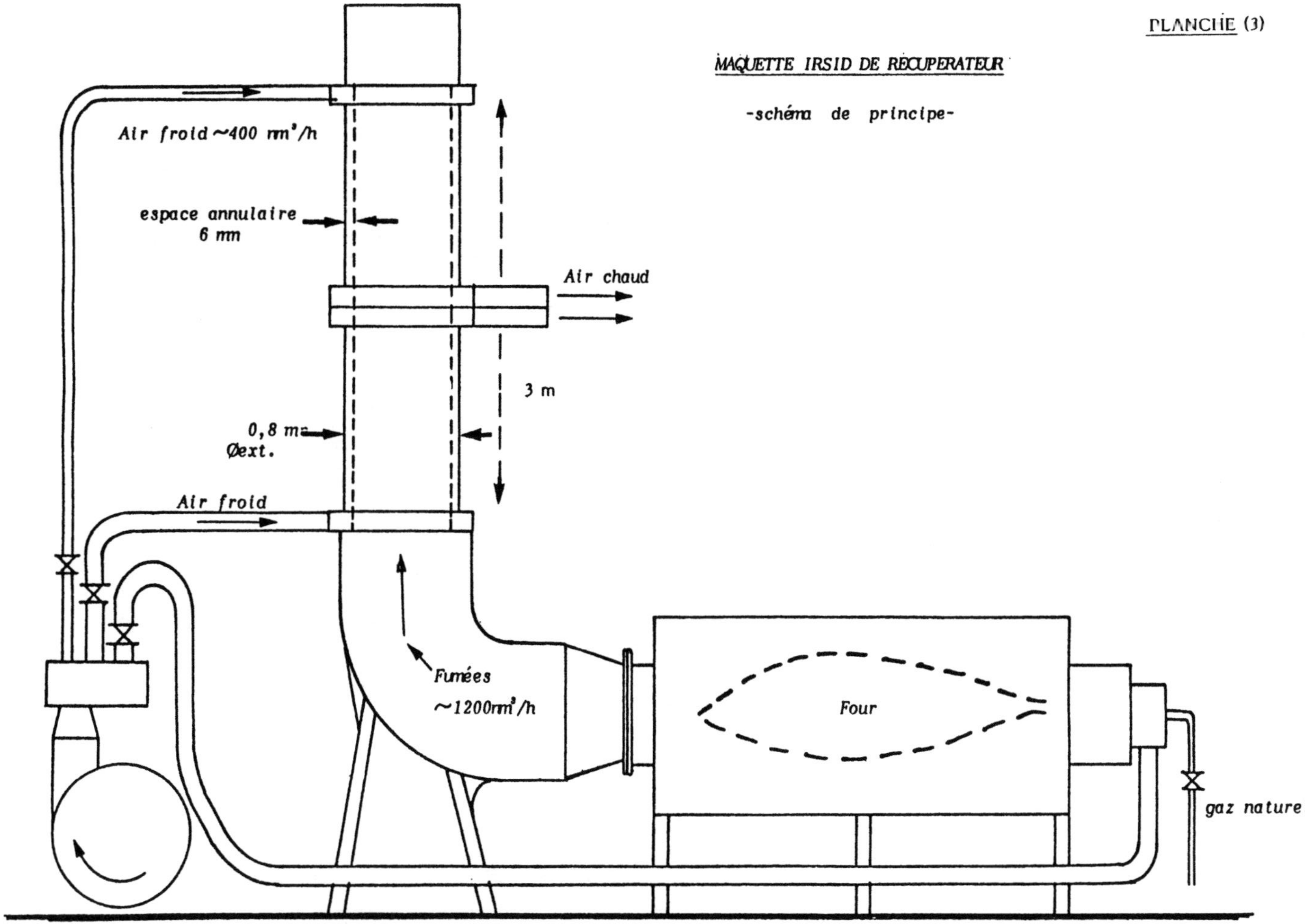

GRADIENTS DE TEMPERATURE SUR UN DIAMETRE DU COURANT DE FUMEES

Planche (4)

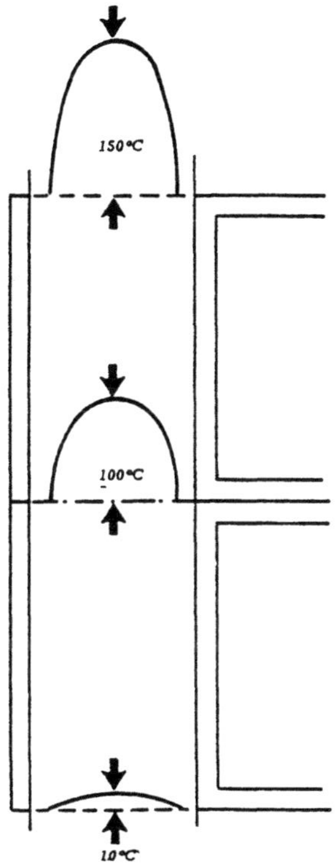

NB: - profils corrigés de la dissymétrie constatée sur la circonférence du récupérateur
- les températures relevées par la mesure fixe sont prises au centre de la conduite

SCHEMA DE PRINCIPE D'UN RECUPERATEUR "ESCHER"

Planche (5)

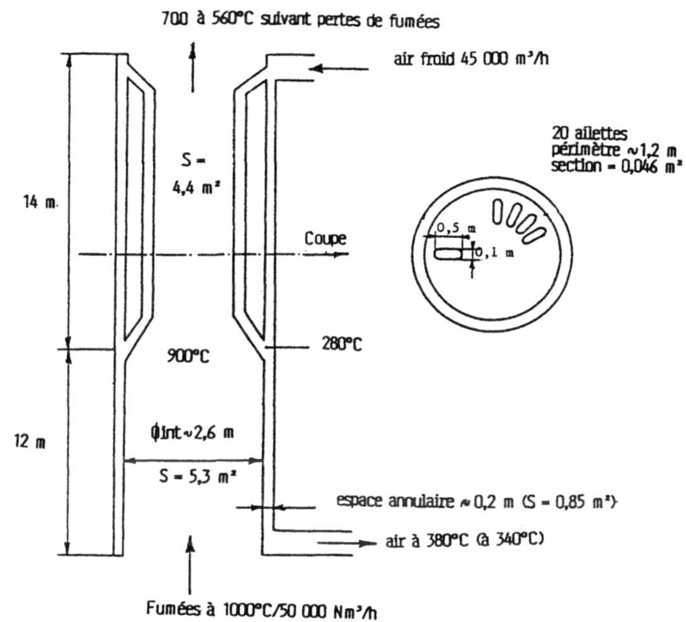

Planche (6) PROFIL DE TEMPERATURE D'UN GAZ EN ECOULEMENT TURBULENT DANS UN TUBE CIRCULAIRE (d'après KNUDSEN et KATZ)

The analogy between momentum and heat transfer

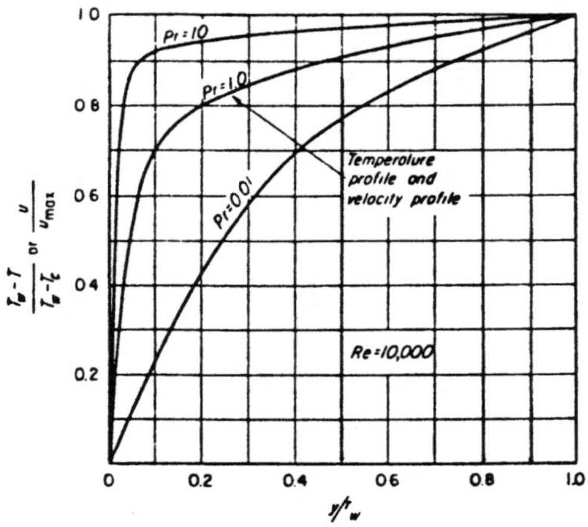

Effect of Prandtl number on temperature profile for turbulent flow in circular tubes. Calculated from Eqs. (15-58) to (15-60).

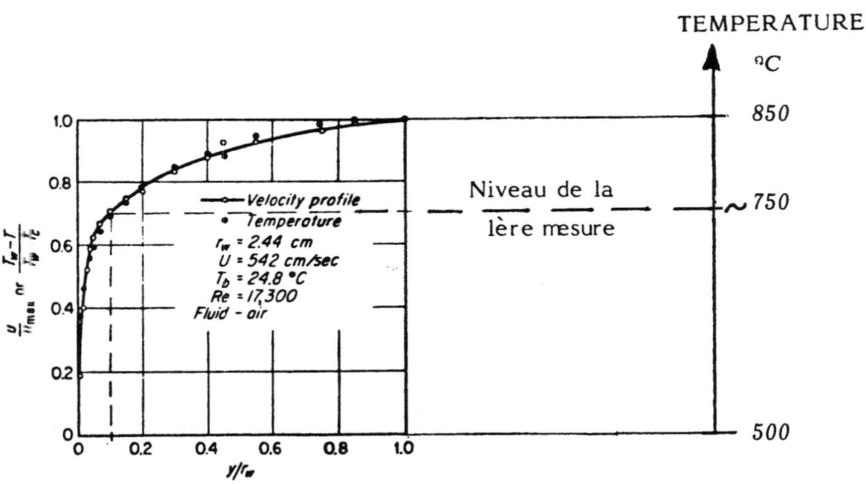

Velocity and temperature profiles for turbulent flow in a circular tube (Data from J.R. Pannell, Brit. Aeronaut. Research Comm. R. & M. 243ii, 1916).

APPLICATION AUX MESURES A MI-HAUTEUR SUR LA MAQUETTE

Planche (7) MISE EN REGIME DE L'ECOULEMENT TURBULENT D'UN GAZ ENTRANT DANS UN TUBE - AUGMENTATION DU TRANSFERT CONVECTIF PAR DES TRUBULEURS

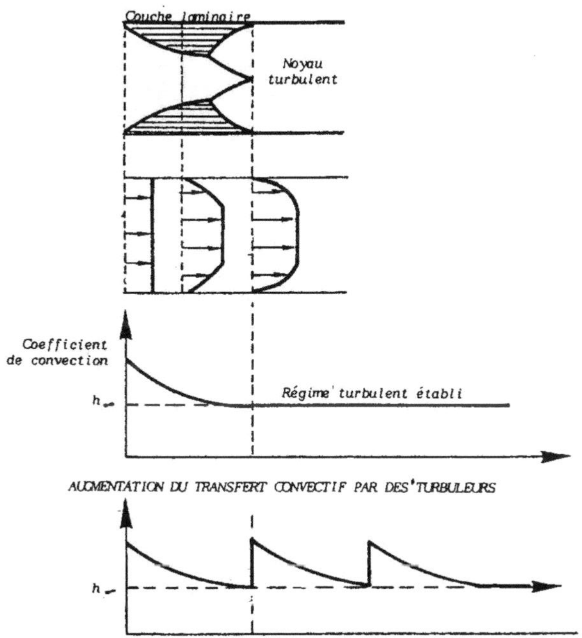

Planche (8) PRINCIPE DES PANNEAUX-RELAIS AUGMENTANT LE TRANSFERT RADIATIF FUMEES → PAROI

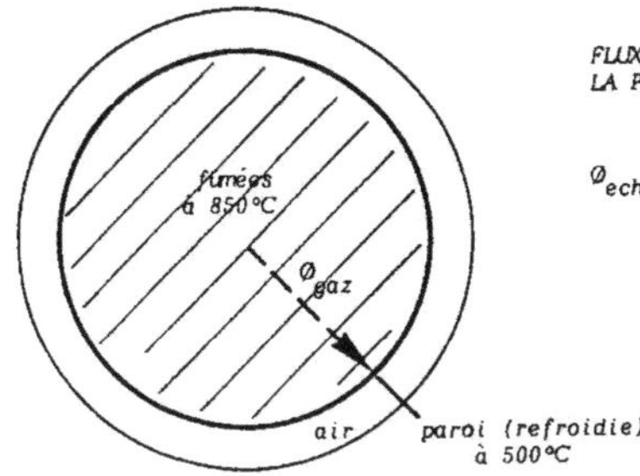

FLUX RADIATIF ECHANGE PAR LA PAROI REFROIDIE :

$$\emptyset_{ech} = \emptyset_{gaz} = \sigma\, \varepsilon_{gaz}\, \frac{1+\varepsilon_p}{2}\, (T^4_{gaz} - T^4_{paroi})$$

$\varepsilon_{gaz} \rightarrow 0,26$

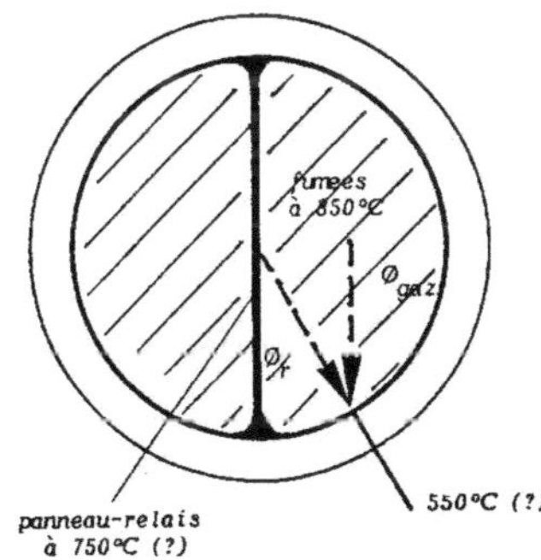

$$\emptyset_{ech} = \emptyset_{gaz} + \emptyset_{relais}$$

$$= \sigma\, \varepsilon_{gaz}\, \frac{1+\varepsilon_p}{2}\, (T^4_{gaz} - T^4_{paroi})$$

$\varepsilon_{gaz} \rightarrow 0,20$ (courant gazeux moins épais)

$$+ \sigma (1 - \alpha_{gaz})\, \varepsilon_r\, (T^4_{relais} - T^4_{paroi})$$

$\alpha_{gaz} \rightarrow 0,3$

Planche (9) DISPOSITIFS SUGGERES POUR AMELIORER LE RENDEMENT DES RECUPERATEURS A RAYONNEMENT

"Cône-turbuleur" entretenant le régime transitoire à l'entrée du récupérateur

"Croisillon-relais" augmentant l'échange radiatif fumées-paroi

'Spirale" · augmentant les échanges radiatif et convectif entre les fumées et la paroi

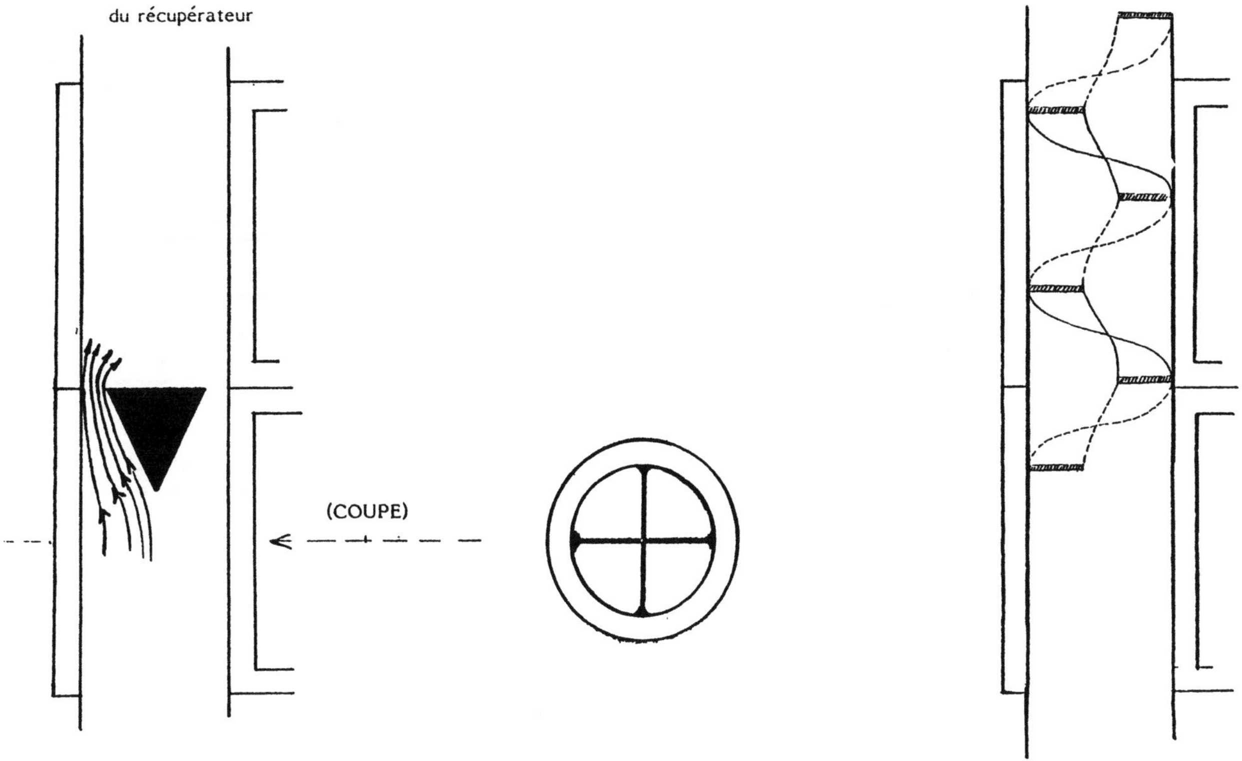

IMPROVED BOILER CONTROL IN FOOD FACTORIES

Author	:	W.E. WHITMAN
Contract number	:	EEB-1-124-GB
Duration	:	30 months 1 May 1981 - 30 November 1983
Head of project	:	W.E. Whitman
Contractor	:	Leatherhead Food Research Association
Address	:	Leatherhead Food Research Association Randalls Road Leatherhead, Surrey KT22 7RY

Summary

The initial stage of this project required : identification of a suitable factory for trials, evaluation of boiler efficiency, analysis of steam loads due to individual plant processes and their timing, development of a control and optimisation strategy and preparation of a costed plan to implement control.

A second stage would be to implement the control strategy if justified by the costing.

A suitable industrial partner PL Group (Foods) Ltd has been identified and a data logging exercise using a computer initiated. Due to specification errors, the important steam flow meters have not been available and the project has been delayed. The steam meters are now operating and a programme of data acquisition and monitoring will shortly commence.

1.1 Introduction

Surveys carried out by the Leatherhead Food RA of energy use in the food industry (1 & 2) have indicated that steam is the most commonly used heat transfer medium used in process heating. Good thermal efficiency in the boilerhouse is therefore very necessary if food manufacture is to be an efficient energy user.

It is the nature of many food manufacturing operations that they are carried out on an intermittent basis. This may be because batch production is the most economic route or because an appreciable time interval is required to complete a manufacturing process. It is commonly the case that more than three quarters of the steam raised is used for process heating and also that the steam requirement may vary by a factor of four over an interval of only a few minutes.

Firms commonly install an excess boiler capacity to ensure continued availability of steam in the event of boiler breakdown and it is often found that a boiler plant runs at a very light load with erratic demand peaks that are well within the capacity of the fired boilers. This pattern of running a boiler is very inefficient. Alternatively, a boiler may be run at a higher loading and then it may be found that a demand surge cannot be met, the steam line pressure falls quickly and this may cause surge conditions in the boilers leading to boilers 'locking out' (3).

Many of these problems would not occur if firms installed steam accumulators or boilers with more steam headspace. These devices are expensive and space consuming and during the post war period of low energy prices their installation could not be justified.

One of the most significant contributions of modern technology to boiler management has been the zirconia oxygen probe that makes possible on line measurement of inferred boiler efficiency. This device can be coupled with a microprocessor to produce a versatile low cost combustion controller. A device of this type has been evaluated successfully in a British food factory (4).

Most food factories have a number of boilers and it was a logical extension to consider the role of the microprocessor in improving energy efficiency in the boiler house as a whole. Two related ideas were apparent. One was to use the computer to provide data on boiler house performance which would make possible better management of the boilers; we are actively pursuing this idea elsewhere. The second idea related to the erratic steam load and low average utilisation of food factory boilers already mentioned.

It was thought that in a factory with a number of steam users, many of which were making intermittent and erratic demands for steam, it would be of great advantage if a forward plan of steam demand could be made. If it were known that ten minutes ahead steam demand would fall, then a number of choices could be available on how to manage the steam distribution system.

This project was proposed to explore the possibilities of making useful energy savings in an actual factory by using suitable sensors and a computer to build up a forward pattern of steam demand a few minutes ahead and to use this with a knowledge of process priorities and boiler efficiency curves to manage the steam generation and distribution system in the most cost effective manner.

1.2 The Factory

PL Group (Foods) Ltd is a rapidly growing company in the edible oils field. It started in business packing lard but now has equipment for most of the standard refinery processes; neutralising, bleaching, hydrogenation and deodorisation, together with blending and packing equipment. Packaged white fats are a growing outlet but the major tonnage is sold as tanker loads of oils. The refinery handles lard, palm, soya and fish oils. The site employs over 200 people with an annual turnover approaching £20 million (5 & 6).

Steam is produced from 3 rather old oil fired boilers which are manually started and have a minimum of instrumentation. The nominal boiler capacities are: No's 1 and 2 8172 kg/hr, No 3 6129 kg/hr at about 16 Bar. Two boilers are required on line to meet normal production demand.

There appear to be five major steam users together with a number of smaller ones. The most important use of steam is to power the vacuum ejectors of the semi-continuous deodorizing plant (the heating requirement is provided by gas). This plant is heavily used and it is important that the ejectors receive a steady steam supply at the correct pressure. Any failure may lead to a loss of vacuum on the plant which not only interrupts production and impairs the quality of the oil being deodorised, but could have even more serious results if water from the condenser system were drawn back into the deodoriser.

A major task of the steam utilisation control programme will therefore be to maintain steam flow to these ejectors as a first priority.

There are two cyclic steam using plants which probably constitute major parts of the demand. One of these is the semicontinuous bleaching plant which is being commissioned to replace much of the batch bleaching capacity. Steam is used to heat the oil to the required bleaching temperature and also to draw a vacuum on the plant. The other cyclic steam using plant is the hydrogenation facility. This is of the Buss loop type and steam is again used, both to draw a vacuum and to bring the oil up to reaction temperature.

The fourth use of steam that is believed to be of major importance is to heat through fats that are brought to the factory wharf by ship or barge. 60,000 tons/year of fats are received in this way and are melted by the use of coils heated by steam from the factory boilers. The Alfa Laval continuous neutralising plants require a steady supply of steam to heat oils to the correct temperature before neutralising with caustic soda and centrifuging to remove soaps and impurities.

The numerous uses of steam that are thought to be of less importance, though in aggregate of real significance, include: remelting fats in the packing plant, heat for the soap splitting and acid oil plant, heating for the extensive tank farm, trace heating for the oil transfer lines, heating for hot water in calorifiers and space heating for offices and workshops.

1.3 The Project Plan

The planned project requires data collection of the size and timing of steam demands from the main steam users with information on the process cycle so that an early signal can be sent to the central computer of changes in steam requirement.

Information is also to be built up on the varying efficiencies of the boilers at different steam loadings. From these pieces of information an action flow chart can be devised that will show how the steam load will be likely to vary in the next few minutes and will indicate the most efficient boiler control strategy to accommodate this variation.

The project envisaged that an estimate of savings to be achieved by improved control could be made relative to the established basic efficiency of the original boiler plant. If the savings justified the necessary capital expenditure then the project would move forward to the major stage of computer control, initially only of the boiler fuel combustion system using an oxygen monitor in the stack gas of each boiler but subsequently the full boiler load control system would be implemented.

Boiler efficiencies would be measured at 3 stages. Firstly at the commencement of the project, secondly when computer combustion control was completed and finally when the full load control system had been commissioned.

1.4 Project Execution

The factory management decided that the most efficient way to gather the necessary information was to purchase a data logging computer for the boiler house operation. This was specified, ordered and delivered at the end of 1981.

With the aid of this computer, boiler efficiencies may be calculated in three ways. An inferred efficiency may be obtained by computing stack gas losses from oxygen and temperature measurements, adding blowdown loss calculated from the blowdown valve setting and making an allowance for radiation and other losses.

An instant enthalpy comparison can be made by comparison of oil and steam flow rates to and from the boiler and a longer term comparison can be made by integrating oil and steam flows over a full shift or other period.

Boiler utilisations may also be computed. A plot may then be made of boiler efficiency against utilisation and by expressing this in money terms the cost of raising steam at any rate can be found.

The computer receives steam flow data from each boiler, from the high and low pressure mains and from four or five steam flowmeters serving main plant areas.

By comparing metered steam flows with the observed operation of the plant, it will be possible to obtain knowledge of the size and duration of steam demand from each standard plant operation and thus to identify for each operation the point in the cycle when the steam demand arises and the size and duration of the demand, together with the major process variables affecting it.

For all this work, efficient operation of the numerous sensors is vital. Unfortunately there have been a number of problems.

The stack gas oxygen probes have had to be significantly modified to give a computer compatible signal, the equipment to measure boiler water conductivity and thus to check 'blowdown' settings of the Gestra valves was on very long delivery and most importantly there were problems with the steam flow meters.

For steam measurement, Neptune vortex shedding meters for a variety of pipe sizes and flow rates were specified but the steam conditions quoted on the orders were unrealistic. Consequently the steam meters as delivered were unsatisfactory and a long delay resulted while the problem was identified, a solution agreed and satisfactory meters and calibrations were prepared and delivered.

The advantages of the vortex shedding meter are that it gives a pulse count that over a wide flow range is proportional to steam flow and the meter is less susceptible to fouling and erosion than an orifice plate. However, it is rather more critical to install it in an adequate length of straight smooth pipe and there is less industrial experience of its application to gas flow measurement. In fact, the only evaluation reports we have seen apply to liquid flow only (7 & 8).

These sensor problems, together with a prolonged breakdown of one of the larger boilers, have severely limited the information we have been able to obtain.

1.5 Observations

Combustion efficiency measurements have been made during 3 visits to the factory using a flue gas sampling oxygen meter and temperature measurement. Some data on the efficiency of No 2 boiler was also obtained when the computer was running for a short time in Dec 1981.

A site energy survey was made in Feb 1981 during which whole site energy statistics for 1980 were analysed.

Due to the lack of steam flow meters, there has been no opportunity to monitor steam loads at individual plants. However some preliminary investigations of potentially major steam using plants have been made as follows:

Semi Continuous deodoriser (SCD); this is a two stream plant that uses gas fired thermex boilers for heating. HP steam is used on the vacuum ejector systems to draw a working vacuum of about 8mm Hg and also to provide steam for direct injection into the oil being deodorised to strip off volatile materials. The steam flow to this plant is separately metered and is the first priority user.

Batch deodorisers; these would be a major intermittent steam user both for heating and to operate the vacuum systems but they are now very rarely used as the SCD can provide the necessary capacity and small parcels of oils are no longer processed.

Semi Continuous bleacher (SCB); this plant is computer controlled and is just being commissioned. Steam will be required to draw vacuum using a single stage ejector and to heat the oil. The load will be of a regular intermittent type to preheat each batch of oil before it enters the bleacher, each batch of oil normally being of the same size and type of oil as the preceding one.

Batch bleachers; the oil charge for these is preheated in a heat exchanger and drawn into the bleaching vessel which is under vacuum. The vacuum ejector is driven by HP steam.

Hydrogenation plant; there are two plants available. Normally only the larger plant is used. If the oil charge is drawn from the batch bleachers, it will be at about 90°C, if it is drawn from the tank farm, the temperature may be only 40°C.

The oil is pumped into a holding tank and then into the heating loop of the Buss loop hydrogenator. Steam is used to heat the oil to 140 or 160°C and then heating is stopped.

A vacuum is drawn on the hydrogenation vessel, to remove air and to draw in the catalyst charge. The hydrogenation reaction is exothermic.

Steam is therefore mainly required for 30 to 60 minutes to heat the oil and not needed during the balance of the reaction cycle of about 4 hours.

1.6 Results and Calculations

Survey of energy use in 1980 (Thrift Visit 15.2.81)

Specific energy ratios 2.56 GJ/tonne
0.454 GJ/employee hour

The boiler house burned 213323 GJ of heavy fuel oil during 1980.

Inferred boiler efficiency:

Boiler	1		2			3		
High fire	80.7	82.5	80.7	81.9	82.2	79.0	81.3	78.5
Medium	76.9	78.8	73.9	78.9	-	73.5	75.9	-
Low	59.3	56.6	60.3	57.4	-	64.5	68.5	-

These readings were taken on different dates which accounts for some of the variation.

In addition, the variation of efficiency of No 2 boiler at different levels of utilisation was calculated from computer printout 11-14 Dec 81 on the basis of GJ steam produced from GJ oil burned. This varied from 75% efficient at 38% load to 84% efficient at 75% load.

Calculated heat requirements for various plants and steam flow rates while heating are:

Plant	Duty	Heat required over 24 hours GJ	Steam flow rate (while heating) GJ/h	Heating frequency
Hydrogenation	Fish oil	6.9	2.3	$\frac{1}{2}$hr every 4h
	Vegetable oil	15.1	3.0	$\frac{5}{6}$hr every 4h
Batch bleacher	Fish oil	10.5	1.2	$\frac{11}{12}$hr every $2\frac{1}{2}$h
Semi Continuous bleacher	Tank farm oil	27.4	2.3	$\frac{1}{6}$hr every $\frac{1}{3}$h
Alfa Laval	2 lines	48.0	2.0	continuous

These are theoretical figures making no allowances for losses etc.

The manufacturer's rating of the steam driven vacuum ejectors on the S.C.D. plant was 1587 kg/hour for each of 2 sets. Thus the two sets together require 6.35 GJ/hour or 152 GJ/day. It is probable that we will find they are using more than that now.

1.7 Future Work

The Neptune meters have now (1.6.82) been recalibrated and refitted on the plant so that steam flow readings can be obtained.

The boiler water conductivity meters are also now fitted and connected to the computer but the oxygen meters still require attention.

A correction is required to the computer programme to take account of the revised duties of the flow meters but it should very shortly be possible to start to build up substantial amounts of data.

There is already an indication that the high efficiency indicated initially for Number 2 boiler may not be substantiated.

Recorder traces of the steam flow to the hydrogenation plant shows the sudden loading that is imposed on the boilers when heating starts.

2. Conclusions

The initial stages of this project required: identification of a suitable factory for trials, evaluation of boiler efficiency, analysis of steam loads due to individual plant processes, development of a control and optimisation strategy and preparation of a costed plan to implement control.

Achievement to date falls some way short of this objective.

The Pure Lard Group (Foods) Ltd factory has been identified as a suitable industrial partner because there was a congruity of ideas on computer control of the boiler house and of the factory steam load. There was a real problem caused by varying steam supply to the ejectors of the deodorising plant, and the process operations contributing to the overall steam demand differed in size and frequency of steam demand with some requirements of much more obvious priority than others.

Monitoring of boiler efficiency has only been partially achieved, as much of the boiler instrumentation was ordered but not commissioned.

In the same way it has been possible to get an idea of process cycles and quantities of oil processed but it has not been possible to monitor steam use.

The beginnings of a control strategy exist but are difficult to complete without more quantitative data on steam users and the supplying boilers.

References

1 Energy Use in the Food Manufacturing Industry. A G Crawford & C R Elson Leatherhead Food RA Sci & Tech Survey 128 (1982)

2 Energy Use in the Meat & Fish and Fruit & Vegetable Processing Industries (MLH214 & 218) to be published by Dept of Industry

3 'The Efficient use of Steam' P M Goodall 1980

4 Comparative Performance of an Oil-Fired 3Pass Economic Boiler with an Advanced Microprocessor Oxygen Control System E.T.S.U. (1981) (Dalgety-Spillers, Wisbech)

5 'The PL Group' Sue Copeland Port of London 1982

6. 'Micros Give Hard Plant a Fat Payback' Process Eng. April 82 p 58-61

7 Evaluation of a Vortex Flowmeter, Evaluation Report No E2252 T75 T.N.O. Holland (SIREP/WIB)

8 Evaluation of a Vortex Flowmeter, Evaluation Report No E2243 T75 T.N.O. Holland (SIREP/WIB)

DISCUSSION

Question - Chairman : What is the state of the programme now ?

Answer We are hoping to start steam measurements on site shortly.

Question - Chairman : Could the results of this work be applied to any factory ?

Answer Each individual factory should be looked at as a special case.

Question - Dr. Pilavachi, CEC : What are your initial ideas on the control strategy ? Would you try to operate the boilers at their highest efficiency ?

Answer The objective of the control strategy is to provide the production processes with the steam flow and pressure they require for efficient operation at minimum energy cost. This may involve increasing or decreasing steaming rate of boilers, cutting steam supply to low priority areas, and restricting supply to some major users in order to maintain supply to priority users. Other options would be available.
We will try to operate the pair of boilers normally required to meet the load at lowest combined running cost - in general this will be the lowest predicted fuel cost to generate the required steam demand.

Question – Dr. Pilavachi, CEC : Did you consider stimulating the system on a computer ?

Answer This is an interesting idea. I am looking into the costs and time required.

CALCULATED STEAM HEATING REQUIREMENTS NEUTRALISING, BATCH AND SEMI-CONTINUOUS BLEACHING AND HYDROGENATION PLANTS.

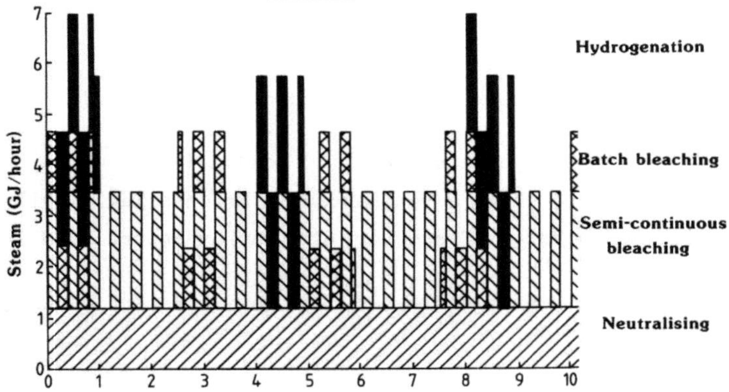

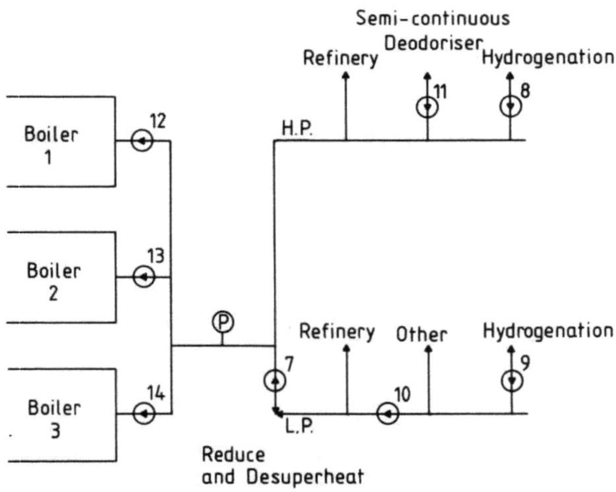

STEAM FLOW MEASUREMENT

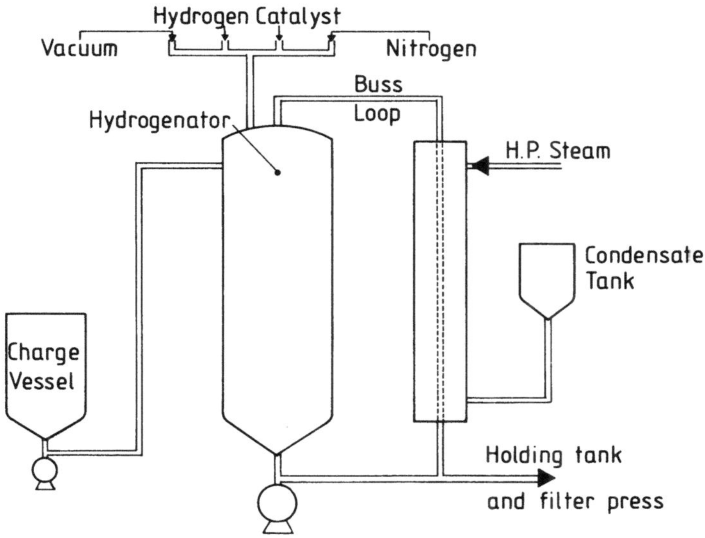

HYDROGENATION PLANT

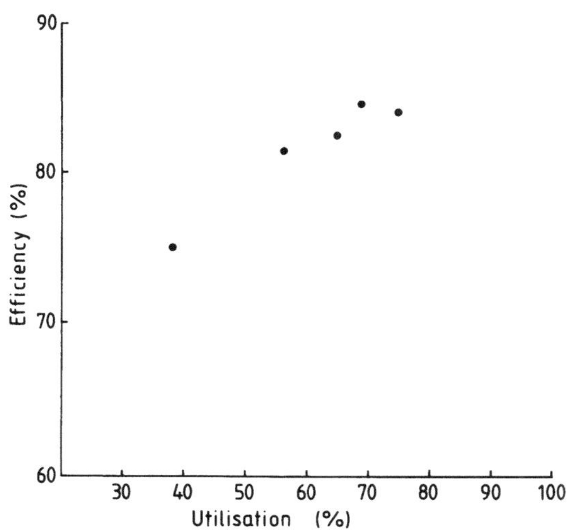

EFFICIENCY (Steam From Oil) P. L. Group (Foods) Ltd

No. 2 Boiler (Dec 1981)

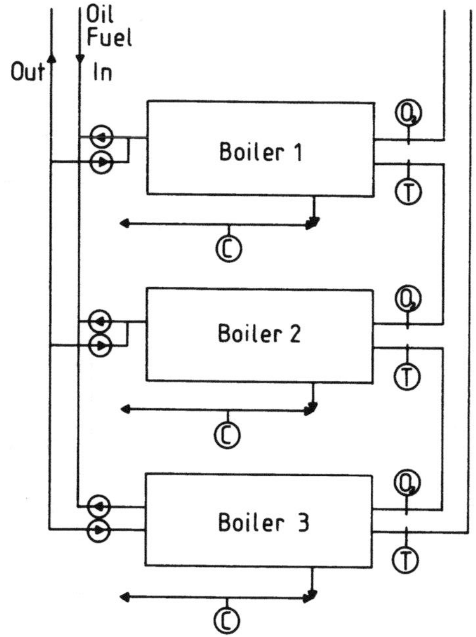

O_2 – Oxygen in stack gas T – Stack temperature
⊖ – Vortex flow meter Ⓒ – Conductivity meter

BOILER INSTRUMENTS

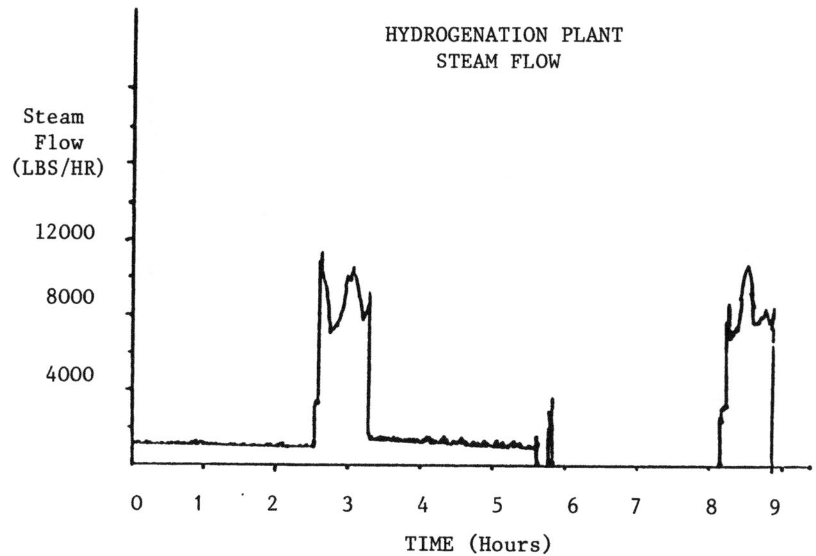

SESSION II - HEAT EXCHANGERS - HEAT RECOVERY

Chairman: D.A. REAY

HEAT EXCHANGERS

HEAT RECOVERY

SUMMARY OF SESSION II

HEAT EXCHANGERS - HEAT RECOVERY

D.A. Reay, P.A. Pilavachi

A large number of industrial processes, covering most industrial sectors, use significant quantities of energy in the form of heat. The efficiency of heat utilization remains such that there is considerable scope for the use of heat exchangers and other forms of heat recovery equipment to enable waste heat to be recycled.

As in the 1st Energy R & D Programme, heat exchangers and heat recovery remained a high priority during the formulation of the 2nd R & D Programme by the Commission, and this is reflected in the scale of the projects now under way, some of which are discussed below. (Note that heat recovery also comes within the project classifications of heat pumps, fluidized beds and Rankine cycle engines). The emphasis within the 2nd R & D Programme is on improved heat exchangers, new materials, and the widening of the operating temperature range of heat exchangers and heat recovery equipment. Successful completion of the projects within these areas will, it is hoped, lead to use of heat recovery equipment in new areas of application. Of particular importance is the growing appreciation by contractors of the need for a system approach to heat recovery, and this is evident from the attention given by contractors to the "peripherals" associated with the heat exchangers under development. *

The success of any research and development project, or programme, may be directly measured by the degree to which the results, if successful, are exploited. This exploitation depends upon many factors, not least of which is the committment of the organisation(s) involved in the work to persue the work. Exploitation can be aided, however, by guaranteeing that the results are given wide publicity, and publishing the proceedings of these contractors' meetings is directed to this end.

The meeting discussed here, as mentioned above, considered papers on heat exchangers and heat recovery. In the context of the Community R & D Programme, the distinction between the two is not great, as all the heat exchanger projects are related to waste heat recovery applications.

Some of the projects highlighted one of the major problems facing potential installers of waste heat recovery equipment, namely the identification of a use for the recovered energy. This will be shown to be particularly true in industries where there is an excess of low and medium grade heat, and electricity generation may be the most satisfactory route to follow.

* In most cases of the research projects, the progress of work has not reached a stage for economical assessment. This would be made at a later stage and economic results would be presented in a future contractor's meeting.

Of possible greater current significance is the fact that changing economic climates affect decisions on investment in waste heat recovery equipment no less than in other areas. The situation persisting at present does not assist the programme in this respect.

Heat exchangers

Of the five papers presented in this session, a majority deal with gas-gas heat exchangers, the exception being an evaporator used in a mechanical vapour recompression (MVR) system. Two papers discuss heat pipes and thermosyphons as components in heat exchangers, (UKAEA HARWELL and the INSTITUT FÜR KERNTECHNIK UND ENERGIEWANDLUNG (IKE), Stuttgart), and BERTIN & CIE describe a ceramic recuperator for use at high temperatures. INSTITUT FRANCAIS DU PETROLE are developing a plate recuperator having the advantage of compactness. COURTAULDS are investigating a complete process, of which the heat exchanger, in this case made of plastic, is only a part.

Heat pipes and thermosyphons - IKE in Stuttgart reported on the three main aspects of its study of two-phase thermosyphons : heat transfer, vapour-liquid flow interaction and materials compatibility.

In stating that while heat pipes were no longer novel, and several forms of heat pipe heat exchanger currently exist, it was pointed out that optimum design and manufacturing procedures were needed in order to enable such heat exchangers to compete effectively with other systems. Furthermore, the complex evaporation and condensation process within heat pipes/thermosyphons was not fully understood, and units operate well below their limits. The final problem area highlighted was that of materials compatibility-this could be demonstrated for vapour temperatures of up to 280° C, but above this the selection of suitable working fluids and container materials becomes more difficult.

Heat transfer coefficient measurements made in the evaporator and condenser showed that while values of 10,000 W/m^2K were achieved in the former, coefficients one order of magnitude greater than this could be measured in the condenser. While the overall conductance of heat pipes was greater than thermosyphons, the throughput of heat pipes was less. Typical axial throughputs achievable, however, are in excess of 1 kW/cm^2.

It was found that high performance thermosyphons were performance limited by liquid/vapour flow interaction, not by boiling.

A major part of the IKE programme is directed at identifying suitable working fluid/material combinations for use in heat pipes and thermosyphons at high temperatures (typically 400° C vapour temperature). While cheap carbon steels may be used as containers for fluids such as naphthalene and toluene at temperatures of up to 270° C and 280° C respectively, more expensive materials are likely to be needed if organics are to be successfully used at higher (350-400° C) temperatures.

An alternative solution to the problem of working fluids for use at temperatures in the range 300-500° C is to select an inorganic fluid. Work carried out at UKAEA HARWELL, following on from an earlier development programme leading to the manufacture of a number of prototype low temperature heat pipe exchangers, is based on the use of sulphur, with added iodine, as the working fluid. The system would operate at moderate vapour pressures, and would therefore not create containment problems in this respect.

The work programme involved studies of both the thermal performance and corrosion behaviour of sulphur/iodine mixtures. Thermal tests on the working fluid in a series of silica thermosyphons showed that predominantly dropwise condensation was occuring in the condenser, but only over the lower (closest to the evaporator) section. The system exhibited characteristics similar to those of a gas-buffered heat pipe, namely a sharp decrease in temperature as the "cold" end of the condenser is approached.

Compatibility tests showed that none of the 6 candidate metals selected as containers would be recommended. Those tested included Hastalloys, Fecralloy and stainless steel.

Ceramic gas-gas heat exchanger - The development of heat exchangers to recover heat in very high temperature processes, where exhaust gases exceed 1000° C, has been attempted by many laboratories. A novel approach has been adopted by BERTIN & CIE, who reported on the progress of their project to develop a gas-gas ceramic heat exchanger for use in exhaust gases of up to 1500° C.

Based on the use of silicon carbide, the design is directed at optimising both convective and radiative contributions to heat transfer. Forced convection is maximised by using jet impingement heat transfer within the ceramic perforated plates of the heat exchanger. This permits overall heat transfer coefficients of the order of 150 W/m^2K to be achieved.

As a result of a detailed market survey, prime uses have been identified in the metallurgical, ceramics and oil (reformers) industries.

Compact gas-gas heat exchanger - While upper operating temperature is one constraint limiting the application of heat exchangers in heat recovery, installation costs, regardless of operating temperature, are also highly important. The project involving collaboration between the INSTITUT FRANCAIS DU PETROLE (IFP) and CREUSOT-LOIRE is directed at the development of a heat exchanger which will assist to minimise installation costs.

IFP has estimated that within the EEC, gas-gas heat recovery could lead to savings of the order of 2×10^6 toe per annum in industrial sectors (excluding glass metals and textiles). The system under development by IFP, which can accept gases at up to 400° C and could have a duty of typically 650 kW, has size advantages over competing systems and its modular construction should also lead to lower installation costs.

Evaporator for Mechanical Vapour Recompression (MVR) systems - A project in which the heat exchanger is a small part of a total process is typified by the work at COURTAULDS in the UK on the evaporation of viscose process liquors. The process involves MVR techniques, in conjunction with the use of thin walled plastic tubes as a heat transfer surface.

A significant part of the development programme is associated with the plastic heat exchanger. The short and long term mechanical stability, resistance to fouling and wettability, are of particular interest. Heat transfer coefficients in the range 800-1000 W/m^2K across the tube bundle have been obtained with the viscose liquors. The evaporation process is directed at the removal of water from the process liquor, and estimated energy savings are of the order of 5×10^8 kWh per annum.

Heat recovery

The session on heat recovery discussed three projects (CERCHAR were unable to present information on their project on heat recovery in coke plant). All these projects deal with metallurgical industry applications. JAMES HOWDEN are studying the use of rotating regenerators on an aluminium melting furnace, a project which originated during the 1st Energy R & D Programme. BRITISH STEEL CORPORATION and TEKSID (a member of the Fiat Group) are both studying energy conservation by heat recovery in the steel industry, where one of the main difficulties is the identification of suitable uses for the recovered energy.

Rotating regenerator on an aluminium furnace - The concept of the rotating regenerator is not new, and it was pointed out by JAMES HOWDEN that about 50,000 such heat exchangers were in use throughout the world. The aim of this project, in which a rotating regenerator would be employed on an aluminium remelting furnace, is to extend the Lungstrom principle to higher temperatures. A maximum of 750° C is specified, giving a preheat temperature of 600° C.

This project has been plagued by the recession in the aluminium industry in the UK. Alcoa, the original partner, scrapped the furnaces where the installation was first proposed. Alcan Plate, the new partner, has cut back production due to low demand from the aircraft industry, and this is delaying work on the current contract.

Waste heat boilers on electric arc furnaces - The iron and steel industry uses a significant proportion of the energy used by industry as a whole, and has received considerable attention from the point of view of energy conservation. Gas-gas heat recovery and pressure recovery techniques are used, but a new field of study is the application of waste heat boilers on electric arc furnaces. It has been estimated that the energy rejected in the exhaust gas from such furnaces could amount to 200 kWh/tonne steel.

In this project, undertaken by the BRITISH STEEL CORPORATION, a number of sites have been studied in detail. Demand for steam varies from site to site, in terms of both quantity and application, and one is also faced with an intermittent and variable heat source. A possible use for a waste heat boiler in combination with, or as an alternative to, a scrap preheater has been identified.

Cast iron melting furnaces - TEKSID, a subsidiary of Fiat in Italy, approached a similar problem to that described above in a slightly different manner, opting directly for the generation of electricity as the end product of the recovered energy. Based on using the exhaust from a cupola, via a waste heat boiler, a turbo-generator plant can, it is predicted, generate 87.5 kWh/tonne cast iron produced.

A pilot plant, to investigate the thermal aspects, has been constructed and successfully operated.

DEVELOPMENT OF A HIGH TEMPERATURE HEAT PIPE HEAT EXCHANGER FOR RECOVERY OF RESIDUAL HEAT

Authors : G. CHAFFEY, J.C. RALPH, C.D. WADE

Contract number : EEB/1/104/80/UK(H) for phase 1.

Duration : 12 months. 1 October 1980 - 30 September 1981 (plus 9 months extension)

Head of project : Mr J.C. Ralph, Engineering Sciences Division

Contractor : UKAEA

Address : UKAEA, Harwell, Oxon, England

Progress paper to be presented at a contractors meeting on "Heat exchangers - Heat Recovery" in Brussels on June 10th 1982.

Summary

Low temperature heat pipe heat exchangers have been successfully developed at Harwell and elsewhere that have been shown to be both efficient and easy to fit to existing plant.

Choice of operating fluids and constructional materials to bridge the high temperature range between about 200°C and 500°C, without resorting to hot side bypass control systems is limited. There is thus a need to investigate any promising-looking new fluids which could possibly be used to fill the present gap in heat pipe technology. The current project is concerned with the latter.

The main part of the programme, which is essentially research in nature, has been concerned with the experimental investigation of sulphur with small additions of iodine as a liquid viscosity reducing agent. Heat transfer tests were carried out on silica and metal heat pipes containing a 5% mixture of iodine in sulphur.

The initial results indicate that the system may have characteristics similar to those of a gas buffered heat pipe.

Results of a series of 1000-hour corrosion tests on some six candidate metals indicates that we cannot at this stage recommend any suitable metal for long periods of exposure to sulphur-iodine mixes at elevated temperatures. It may be noted, however, that silica was unaffected by contact with the mixture.

We hope to investigate a number of other materials before the end of the project and to further study the heat transfer characteristics of the system.

1. Introduction

Low temperature heat pipe heat exchangers have been successfully developed at Harwell for waste heat recovery and have been shown to be both efficient and easy to fit to existing plant. The system was based on the use of copper as the construction material of the pipe and external fins, with water as the working fluid. Such systems are limited both on strength and oxidation grounds to temperatures below about 150°C. This rules out the use of these heat exchangers for a large fraction of the energy recovery market.

The potential savings possible in the temperature range from about 200°C to 500°C appear to be greater than for temperatures both below and above this and methods of recovery in this mid-temperature zone are thus worthy of particular attention.

Choice of operating fluids and constructional materials to bridge the range between about 200°C and 500°C is limited to a few organic derivatives typical of which are diphenyl compounds, and mercury. The organic derivatives decompose to varying degrees above about 300°C and mercury cannot seriously be considered for widespread commercial use on the grounds of both cost and toxic hazard. There is thus a need to investigate any promising-looking new fluids which could possibly be used to fill the present gap in heat pipe technology. The current project is concerned with the latter.

The main part of the programme which is essentially research in nature, is to assess the heat transfer performance of a new candidate fluid based on sulphur plus additives. A recent report [1] has indicated that the unacceptably high viscosity of pure liquid sulphur can be significantly reduced by the addition of a few percent of a material such as iodine.

It is likely that heat pipe corrosion problems will be the limiting factor in the use of sulphur-iodine mixtures for practical high temperature heat pipe heat exchangers. Thus an important part of the programme is concerned with testing possible materials for resistance to surface corrosion by sulphur-iodine mixtures.

2. Description of Research Activities

2.1 Design and construction of basic test facility

Before starting any detailed design or construction of actual heat pipes, an assessment was made of the necessary basic test facilities required. Having established the general nature of the heat transfer and corrosion tests to be undertaken, an early start was made on the construction of a number of furnaces, control systems, vacuum plant and safe working enclosures for the systems.

The test plant, which was designed and constructed specifically for the project, consists of four temperature controlled vertical tube furnaces for corrosion testing of potential heat pipe materials at temperatures up to 600°C, plus several similar facilities for heat transfer tests. A vacuum plant was also provided for the preparation of the working fluids and for use during the filling of test pipes.

2.2 Physical properties of working fluids

Part of our initial preparatory work was, of course, concerned with the acquisition of the relevant physical property data for the most promising-looking candidate fluids.

Since organic derivatives such as "DOWTHERM*" and "THERMEX+" have

* DOW Chemical product name
\+ ICI Product name

already been used in heat pipes designed to operate at several hundred °C, it was initially planned to investigate any new and promising-looking additions to the commercially available range of fluids of this type. However, subsequent contact with manufacturers and specialists has revealed that all presently known organic derivatives are likely to decompose at unacceptable rates above about 300°C and can thus only be used in gas streams above this temperature if a hot gas bypass system is provided to take care of loss of heat sink cases.

However, research elsewhere [1, 2] had indicated the promise of systems based on sulphur plus additives and, because of the relatively short timescale of the present study, we are concentrating our main effort, initially, on this system.

Properties of sulphur plus additives. It was suggested many years ago [2] that provided a stable additive could be found to reduce the anomolous dynamic viscosity of sulphur the latter had considerable potential as a heat transfer fluid.

The liquid viscosity and vapour pressure of pure sulphur is shown in Fig. 1 and 2 respectively. It is clear from Fig. 1 that the viscosity of liquid sulphur as a function of temperature exhibits a rather unusual characteristic. After an initial small reduction just above the melting point (~ 112°C) the dynamic viscosity rises rapidly by some four orders of magnitude to a maximum around 100 Ns m^{-2} at about 190°C, subsequently falling again to the original minimum of about 10^{-2} Ns m^{-2} at around 600°C.

The modern explanation [3] of this anomolous behaviour is that near the melting point the sulphur, which is predominantly at this stage S_8 in ring form, starts to break down forming chains of molecules. The interaction of these polymer chains increases as the temperature is raised and is responsible for the rapid rise in viscosity seen up to about 190°C. Beyond this temperature progressive breakdown of the molecular chain structure occurs, causing a drop in viscosity. The thermal conductivity of liquid sulphur stays almost constant at about 0.16 W m^{-1} $^{o}K^{-1}$ over the range of interest.

It has been recognised for some time [2, 4, 5] that trace impurities in sulphur can bring about dramatic changes in the liquid viscosity, and in particular the addition of halogens to sulphur was shown [2] to offer the possibility of a number of stable systems.

The long-term stability of iodine, bromine and chlorine additions as viscosity-reducing agents for sulphur was investigated by Bacon and Fanelli [2] who were optimistic about the results. More recently, Polasek and Stulc [1] have also used sulphur and iodine mixtures for heat pipes, and have measured the viscosity of this system. The results of this work are summarised in Fig. 3 where the viscosity of 3, 5 and 10% additions of iodine to sulphur are shown as a function of temperature.

The corresponding effective vapour pressure of the iodine-sulphur mixtures are summarised in Fig. 4. These are theoretical estimates and are likely to be in error by about 20% because of the relatively complex nature of the system.

2.3 Heat transfer tests with sulphur plus iodine

2.3.1 Silica heat pipes

In order to evaluate the heat transfer performance of the selected working fluid, several silica heat pipes have been constructed. The use of silica for initial trials offers several advantages over metal; it is possible to observe the condenser section, the pipes are relatively cheap and quick to manufacture and are highly resistant to chemical attack.

The arrangement of the silica pipes is shown in Fig. 5. They are 0.96m long with a bore of 0.03m and have a condenser, which is air-cooled, 0.6m long. The evaporator section, 0.3m long, is heated electrically using a wrapped-on co-axial heater cable.

Thermocouples are attached to the outside of the 1.7mm thick heat pipe wall using a high temperature ceramic-based cement. Deep thermocouple pockets were provided in the ends of the first pipe in order to measure the evaporator liquid temperature and the condenser vapour temperature. However, the evaporator pocket was omitted on subsequent pipes after early fracture of the pocket occurred due, it appears, to volume changes of the sulphur-iodine mixture under test during solid-liquid phase change. The pocket was replaced by a shallow cavity in the base of the evaporator which is not directly in contact with the heater cable and would thus be expected to indicate a temperature close to that of the surrounding evaporator liquid pool. A further modification to the silica pipes, four of which have been made to date, is the provision of a vent line at the top of the condenser to allow for possible venting of small quantities of trapped non-condensables which may be present.

The technique adopted for filling the first heat pipes was the relatively straightforward one of melting the sulphur and iodine together in a covered glass container (vented to atmosphere) and pouring the melt directly into the pipe.

The sulphur used was sublimed commercial grade and the iodine analytical grade, a 5% by weight mixture of the latter being prepared by weighing of the two constituents prior to melting. To minimise loss of the relatively volatile iodine during the melting process, the crystals were placed between two layers of sulphur powder in the melting vessel.

The amount of mixture placed in the heat pipe occupied about 68ml and was based on the recommendations in the literature [6] for optimum fill. The filled heat pipe was cooled and joined to a vacuum system, and several sequences undertaken of:- pumping; sealing off; melting to out gas; re-cooling to liquid nitrogen temperature, followed by a further period of cold pumping. The vacuum reached during the pumping stages was approximately 10^{-4} TORR, which was also the level of vacuum during the final sealing-off sequence.

The prepared heat pipes are fitted with heaters, lagged and then placed in position in the test facility.

A series of heat transfer tests has been carried out using silica pipes of a general layout described above. The results from these have indicated the special precautions required when preparing the working fluid and the somewhat unique behaviour of the system when in operation.

Observation of the condenser during operation revealed that condensation appeared to be only taking place over the lower part of the cooled section. Filmwise condensation was observed but only for early low power runs. For the majority of runs the condensation was dropwise in character as shown in Fig. 6.

Some results typical of our observations on silica pipes are shown in Fig. 7.

The results indicate that the system has characteristics similar to those of a gas buffered heat pipe in which the condenser is not all at uniform temperature but varies its effective length as well as temperature with power.

The source of this gas if present can be residual gas due to inadequate vacuum before sealing, outgassing occurring after sealing, or by the iodine vapour functioning as an inert buffer. If the buffering is caused by poor vacuum conditions during construction or by outgassing, venting the inert gas while the heat pipe is running should expel the inert gas permanently. If the buffering is caused by partition of the iodine from the vapour, venting the pipe will enable the whole of the condenser to function temporarily. A silica heat pipe was prepared as described previously except that instead of sealing the pipe the top of the condenser was fitted with a tap. The heat pipe could be vented into a volume capable of being evacuated and isolated. The heat pipe was run in the normal way (eg. electrically heated evaporator and air cooled condenser). When equilibrium conditions were attained, the heat pipe was vented into the evacuated volume. As soon as the vapour front was observed at the top of the heat pipe the isolation valve was closed. After a few seconds the "buffered" volume again reappeared. The pipe was vented several times, always with the same result. The result was not entirely conclusive however since the vapour temperature was not measured in the silica heat pipes and simple overcooling of the condenser may have accounted for the noted behaviour.

2.3.2 Metal heat pipes

The determination of accurate heat transfer coefficients for the evaporator and condenser was not practical with the silica heat pipes. The thermal conductivity of the silica is low (and somewhat uncertain). Efforts to obtain accurate vapour temperatures have been frustrated by breakages of the thermocouple pockets and heat flux variation down the length of the condenser from the cold air inlet to the relatively hot outlet also aggravates the situation.

To obtain improved results it was decided to build a metal heat pipe designed to operate with an approximately isothermal condenser, the latter being achieved by the use of water cooling and a variable conductivity static gas gap. The thermal conductivity of the gap could be varied by changing the composition of the filler gas from pure helium to pure argon with mixtures of helium and argon if necessary. The vapour temperature is measured by thermocouples mounted in a thin wall tube located coaxially down the centre of the heat pipe. Evaporator and condenser outside wall temperatures were measured by thermocouples fixed to the walls, Fig. 8 shows a diagram of the layout.

The choice of material for construction of the metal heat pipe was difficult since to avoid delay in the relatively short research programme manufacture of the heat pipe had to be started before any corrosion tests were completed. There was some evidence in the literature that stainless steel to an East European specification (CSN 17246) was corrosion resistant[1]. This steel has a specification of 17-20% Cr, 8-11% Ni, Ti stabilised.

A quick corrosion check was carried out on a similar readily available steel by suspending a bar of 18/9/1 S/S in a quartz heat pipe. This can be seen in Fig. 6 as a dark line down the centre. Visual examination after a few hours at ~ 300°C did not show any severe corrosion, therefore construction was carried out using 18/9/1 stock material.

The first low power (250w) heat transfer run on the metal pipe indicated a drop off in vapour temperature about half way alone the condenser with a corresponding drop being noted on the outside wall, venting did not significantly modify the temperature profile which is shown in Fig. 9. However, increase in power to 0.5kw produced stable operation with no indication of behaviour associated with gas or vapour buffering. The 0.5kw result is shown in Fig. 10. It is immediately apparent from Fig. 10 that the condensation heat transfer coefficient is significantly less than that in the evaporator. The calculated average values of the coefficients for this run were 153 $wm^{-2}\,°C^{-1}$ and 1329 $wm^{-2}\,°C^{-1}$ for the condenser and evaporator respectively.

Further increase in power to lkw produced intermitted sharp jumps in both wall and vapour temperature part way along the evaporator region which could only be interpreted as indicating dryout.

The temperature profile along the pipe for several scans at a fixed power of lkw is shown in Fig. 11. Each scan of temperature took 5 minutes and some ten scans are shown. The progressive increase in temperature level with time appeared to indicate degradation of the pipe wall and after a further series of similar runs which showed further deterioration it was decided to open the pipe and examine the interior conditions. This was done and a considerable amount of flakey corrosion product was found on the walls. This came as no real surprise since by this time we had completed a series of corrosion tests (which are described below) which showed that the steel used to construct the heat pipe corrodes quite seriously in the sulphur iodine mixture used.

2.4 Corrosion tests

Samples of materials selected by reference to the literature were exposed to a sulphur + 5% iodine environment.

Coupons of Hastalloy B2, C276 and Haynes alloy 188 were supplied by Cabot alloys of England. These coupons were 30 x 10 x 3 mm. Samples of 18/9/1, 25/20 stainless steel Fecralloy and low carhon steel were obtained from standard stock material and cut into 20 x 10 mm coupons. The mild steel and 20/25 were approximately 1 mm thick, the Fecralloy and 18/9/1 approximately 0.5 mm thick. Typical chemical analysis of materials used are given in Fig. 12.

Coupons were thoroughly degreased, washed in demineralised water and weighed before sealing into glass tubes. Each tube contained two samples separated by a constriction. One end of the tube was fitted with a pumping stem through which weighed amounts of sulphur and iodine could be admitted prior to attaching to a vacuum system. Once attached to the vacuum system the tube was cooled with liquid nitrogen, evacuated and isolated from the vacuum system. The sulphur/iodine was heated until it melted, cooled and repumped. This procedure was repeated several times to remove non-condensable gas while minimising the removal of iodine vapour. The tube was then sealed off. Each tube containing corrosion samples and sulphur/iodine was enclosed in a thin wall stainless steel tube (to prevent multiple breakage in the event of one tube exploding) and heated in a vertical furnace at the test temperature. Each furnace contained a bundle of seven tubes. The lower sample in each tube was exposed to the sulphur/iodine liquid, the top sample to the vapour.

The temperature of exposure were 200°C, 400°C and 600°C. The 200°C and 400°C capsules were made from pyrex glass, the 600°C from silica.

After four hours exposure at 600°C the capsule containing Hastalloy B2 broke, tripping the furnace off. The coupons of B2 were recovered and the remaining undamaged six capsules exposed for another 108 hours. It is believed the capsule containing B2 broke because of the build-up of corrosion product. The samples at 200°C and 400°C were exposed for 1000 hours.

After exposure and cool down the tubes were cracked open and the corrosion coupons removed. The top specimens were loose in the sample tubes, but the lower specimens were set in solid sulphur/iodine, and were removed by melting the sulphur/iodine mixture. After removal all the specimens exposed were photographed, the top specimens as removed, the bottom specimens after minimal cleaning by brushing or very gentle scraping. Adhesion between the scale and the corrosion coupon was low, the corrosion product coming off to reveal the metallic surface in the case of the Hastalloy C276 and B2, a substantially clean surface with some tarnishing in the case of the Haynes Alloy 188, and a uniform dark grey colouration on the surface of the Fecralloy and stainless steels. The mild steel was completely corroded at 400°C and 600°C, at 200°C the corrosion product came off very easily exposing bare, shining metal. The coupons exposed to the sulphur/iodine liquid looked similar to coupons exposed to vapour once the sulphur/iodine had been removed, except that in most cases some of the corrosion product fell away while the sulphur/iodine was being removed.

The corrosion produce was carefully removed from the lower coupons and the coupons weighed. Weight losses are recorded as weight loss/cm² in Fig. 13.

Since only relatively limited corrosion testing was possible with the timescale of the present research contract, reaction rates were not determined but the following general observations have been made.

At 200°C Hastalloy B2 and Hastalloy C276 have good corrosion resistance, but high corrosion rates at 400°C and 600°C. These materials have high nickel content. It is possible that the corrosion film produced at 200°C is protective but at 400°C and 600°C it is not. This could support the recent findings(3) that sulphur reacts with nickel, yielding a low melting point Ni-NiS eutectic.

Haynes alloy 188 and 25/20 stainless steel have better corrosion resistant properties than the other specimens despite fairly high nickel content. The element they have in common in large amounts is chromium (although the major constituent of Haynes Alloy 188 is ~ 35% cobalt).

Fecralloy, although nickel free, does not appear to have particularly good corrosion resistance despite containing 15% chromium. It is much better than mild steel, not as good as 18/9/1 stainless steel at 400°C but better at 600°C.

The corrosion of all the specimens is greatly increased as the temperature increases. Corrosion after 1000 hours at 400°C is an order of magnitude greater than at 200°C. After 100 hours at 600°C the corrosion is more than an order of magnitude higher than corrosion after 1000 hours at 400°C.

Very tentative conclusions would suggest materials based on high chromium content without iron or nickel may be corrosion-resistant.

The observations of Fanelli(1) and Poleseck and Stule(2) that stainless steel is corrosion-resistant cannot be substantiated.

It is worthy of note that all the corrosion experiments and heat pipes used in the programme to date that have been contained in glass or silica have shown no corrosion of the containment walls.

3. Future work

3.1 The sulphur/iodine system is the only fluid, apart from mercury, directly usable in the temperature range 300°C - 500°C. Corrosion represents a major problem, and it is suggested that further work is required including looking at the existing coupons exposed to sulphur/iodine vapour in more detail to try and gain further understanding of the corrosion mechanism, and to follow up a recent suggestion(3) that titanium may be corrosion-resistant.

3.2 Heat transfer data on the sulphur/iodine system is sparse. The modified design of heat pipe we have constructed from 18/9/1 stainless steel has been used to obtain some heat transfer for a limited time. Further heat transfer measurements should be done on metal heat pipes when a compatable containment material is found.

3.3 The observations we have previously made, that all sulphur/iodine pipes we have made appear buffered, should be checked carefully.

The experiment with a vented silica pipe and the metal pipe may have produced misleading observations because the corrosion product from the 18/9/1 stainless steel produced a gas buffer (the low melting point Ni NiS eutectic).

An application to the UK Department of Energy has been made for sufficient funds for completion of the above programme and a final report on the project has thus been postponed in anticipation of further results.

Acknowledgement

The authors wish to thank Mr N J M Wilkins and Dr M H Rand of Harwell's Materials Development Division, for their contributions on corrosion and physical properties of the sulphur-iodine system.

References

[1] Heat Pipes for the Temperature Range 200°C to 600°C, F Polasek and P Stulc, Proc 2nd Int Heat Pipe Conference, Bologna, 1976 (ESA SP 112, Vol 2, pp 711-732).

[2] Bacon, R F and Fanelli, R, "The Viscosity of Sulphur", J American Chem Soc 65, April 1943.

[3] Elemental Sulphur, Ed Beatmeyer, Pub Inter Sciences, 1965.

[4] Farr, C C and Macleod, D B, "On the Viscosity of Sulphur", Royal Soc Proc 1920, 97.

[5] Farr, C C and Macleod, D B, "Some Physical Properties of Gas-free Sulphur", Proc Royal Soc, 118, Series A, 1928.

[6] Strel'tsov, "Theoretical and Experimental Investigation of Optimum Filling for Heat Pipes", Heat Transfer Soviet Res., Vol 7, No 1, pp 23127 (Jan 1975).

DISCUSSION

A point of interest brought out in the presentation of this paper was the mode of condensation in the silicon heat pipes. The large temperature gradients evident in the condenser section also aroused comment.

Question - Dr. Pilavachi, CEC : You mentioned dropwise condensation. Did you investigate the effect of iodine concentration or of an additional component on the heat transfer coefficient ?

Answer The condensation mode in the silica pipes was mostly dropwise. We have only used iodine as an additive and as yet we have not used more than one mixture, although such tests were planned.

Question - Dr. Groll, IKE : Could you observe the type of evaporation (nucleate or film boiling) in the silica heat pipe ?

Answer No. The heater obscured the view of the evaporator section. Perhaps we can arrange a system to do this during future tests.

Question - Dr. Spendel, IKE : What is the reason for the low condensation heat transfer coefficient ? Is it more associated with the presence of non-condensible gases, or is it due to the type of condensation, e.g. mostly film condensation and only partly dropwise condensation ?

Answer On the evidence so far, it would seem that the reason for the relatively poor condensation coefficient is associated with the basic thermal properties of the condensate. We observed mostly dropwise condensation in the silica heat pipes, but do not know what mode occurred in the metal pipes. Corrosion products may also have had an effect in the case of the latter.

Question - Mr. Galant, Bertin : What about using cheap glass ceramics instead of metal tubes ? Do you have any experiments using these and what are the industrial prospects ?

Answer We know that silica is compatible with iodine-sulphur mixtures and have considered that there may be interest in special cases for applying this fact. We have only tested single tubes, not a full heat exchanger.
The prospects for industrial scale glass or ceramic heat pipe heat exchangers must, at this stage, I think be fairly low. More conventional ceramic heat exchangers offer more prospect because some cross-flow leakage can be allowed. Heat pipes are, however, a very different matter; they are pressure vessels and no leakage at all can be tolerated.

Question : Is there an explanation for the large temperature difference along the condenser section ?

Answer The heat transfer coefficient associated with the condenser is 150 W/m^2K, and the evaporation coefficient is 1500 W/m^2k. No change in operating conditions was noted after venting the pipes. This may be associated with condensation phenomena.

Question - Mr. Robertson, B.S.C. : The British Steel Corporation was interested some years ago in high temperature heat pipes for cooling blast furnace tuyeres. What are safety and temperature limitations on heat pipes ?

Answer Provided that the correct fluid, probably an alkali metal, was selected, there is every reason to believe that such a system would operate safely and efficiently.

Figure 1

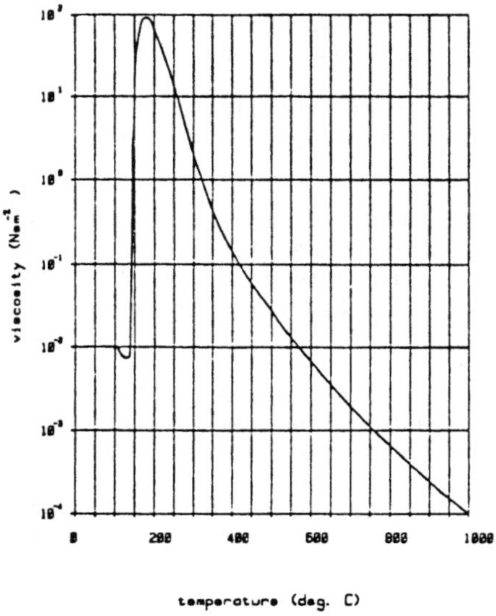

VISCOSITY OF SULPHUR

Figure 2

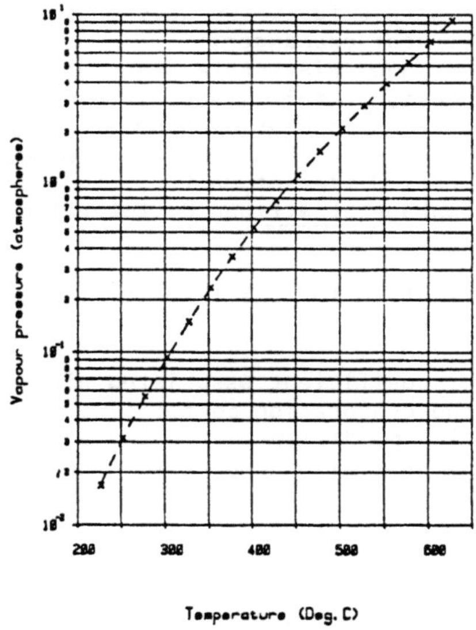

VAPOUR PRESSURE OF SULPHUR

Figure 3

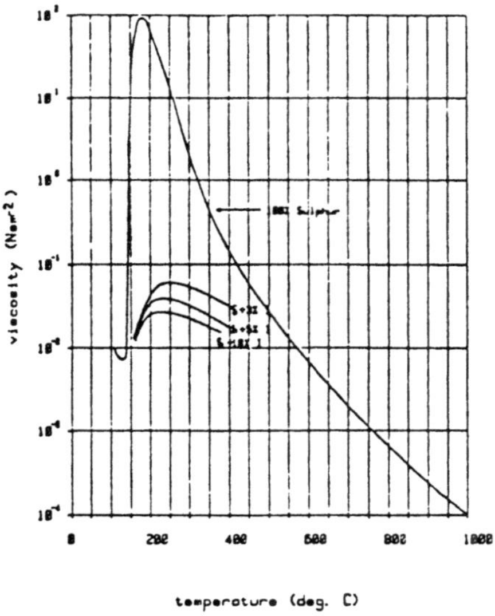

VISCOSITY OF SULPHUR
AND SULPHUR + IODINE

Figure 4

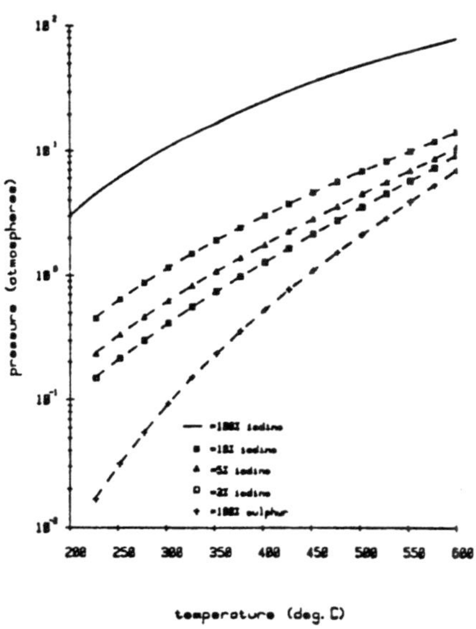

vapour pressure of sulphur, iodine
and sulphur/iodine mixtures

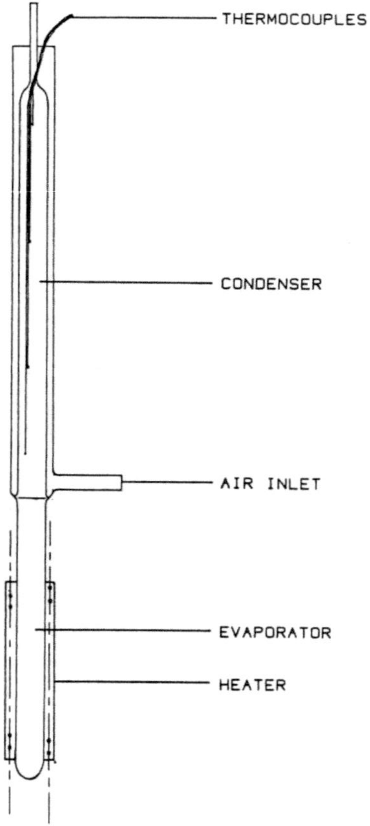

Figure 5

SILICA HEAT PIPE

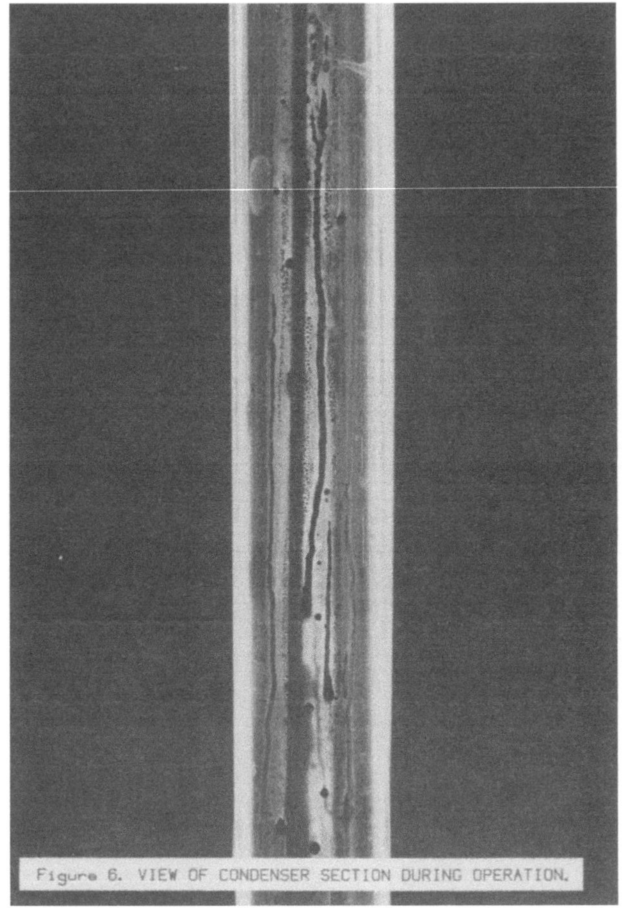

Figure 6. VIEW OF CONDENSER SECTION DURING OPERATION.

1st. Sulphur/5% Iodine Heat Pipe
Temperature Distribution

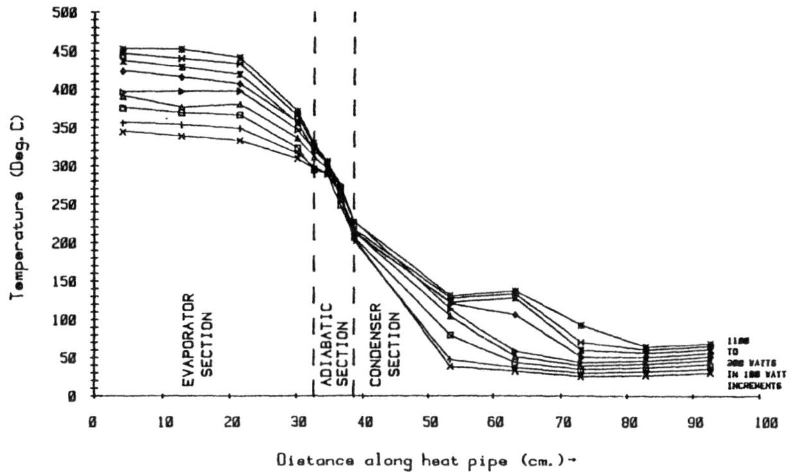

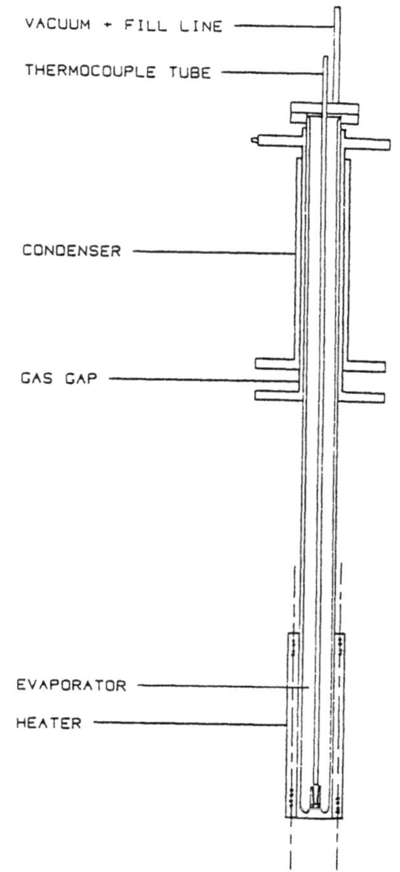

Figure 8

METAL HEAT PIPE

1st. Metal Heat Pipe; Sulphur+5% Iodine

Temperature Distribution

Figure 9

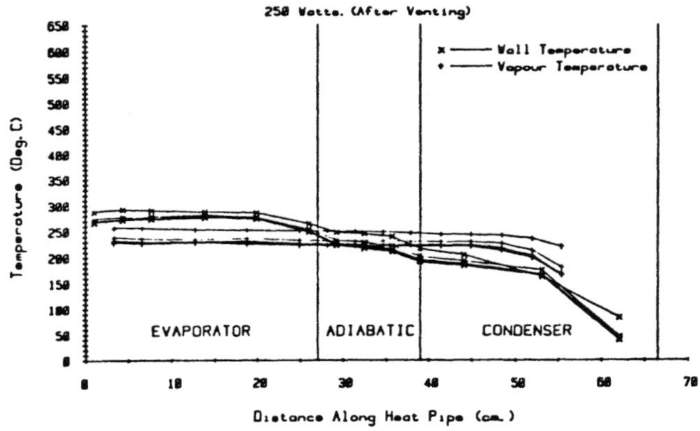

Figure 10

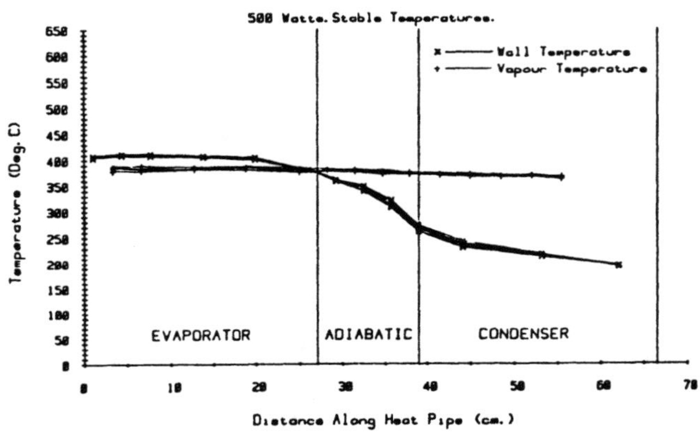

Figure 11

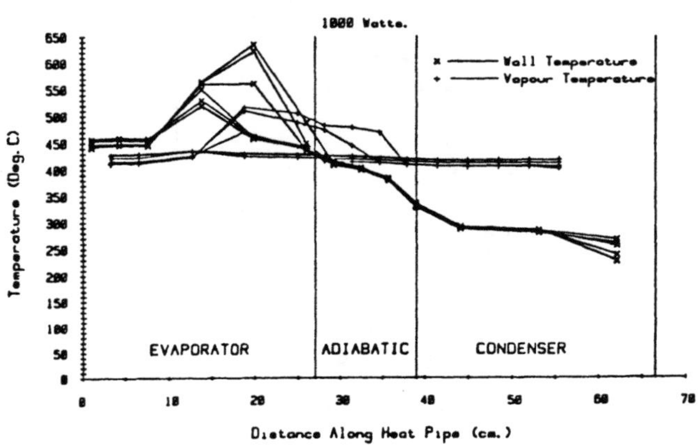

TYPICAL CHEMICAL ANALYSIS

MATERIAL	C	Co	Cr	Fe	Mn	Mo	Ni	P	S	Si	W	Other #
HASTALLOY B2	0.02•	1.0•	1.0•	2.0•	1.0•	28	BAL.	-	-	0.10•	-	-
HASTALLOY C276	0.02•	2.5•	16	5	1.0•	16	BAL.	-	-	0.08•	4	V 0.35•
HAYNES ALLOY 188	0.10	BAL.	22	3.0•	1.25•	-	22	-	-	0.35	14	La 0.04
18/9/1 BS.1449/Pt2 Type 321.S12	0.08•	-	[17.0 19.0]	BAL.	[0.50 2.0]	-	[9.0 12.0]	0.045•	0.03•	[0.20 1.0]	-	Ti 5C/0.70•
25/20 BS.1449/Pt2 Type 310.S24	0.15•	-	[23.0 26.0]	BAL.	[0.50 2.0]	-	[19.0 22.0]	0.045•	0.03•	[0.20 1.0]	-	-
FECRALLOY	0.029	-	15.6	BAL.	-	-	-	-	-	0.23	-	[Al 4.99/Y 0.32 La 0.022]
MILD STEEL Type CR4 BS.1449/Pt1	0.12•	-	-	99.10	0.60•	-	-	0.05•	0.05•	-	-	-

• INDICATES MAXIMUM %
C INDICATES CARBON CONTENT

Figure 12

WEIGHT LOSS per UNIT AREA

MATERIAL	AREA Cm^2	WEIGHT LOSS gm/cm^2 200°C [illegible]	400°C [illegible]	600°C [illegible]
HASTALLOY B2	8.40	7.0×10^-[illegible]	5.4×10^-[illegible]	4.3×10^-[illegible] *
HASTALLOY C276	5.60	7.0×10^-[illegible]	2.8×10^-[illegible]	3.2×10^{-1}
HAYNES ALLOY 188	8.40	1.5×10^-[illegible]	7.8×10^-[illegible]	1.0×10^{-1}
18/9/1 BS.1449/Pt2 Type 321.S12	4.33	2.8×10^-[illegible]	1.3×10^-[illegible]	-
25/20 BS.1449/Pt2 Type 310.S24	4.60	3.2×10^-[illegible]	7.2×10^-[illegible]	1.0×10^{-1}
FECRALLOY	4.45	2.1×10^-[illegible]	2.0×10^-[illegible]	1.2×10^{-1}
MILD STEEL Type CR4	4.60	2.1×10^-[illegible]	-	-

* capsule broke after approx 4 hrs.

Figure 13

HEAT TRANSFER, VAPOR-LIQUID FLOW INTERACTION AND MATERIALS COMPATIBILITY IN TWO-PHASE THERMOSYPHONS

Authors : M. GROLL, O. BROST, D. HEINE, TH. SPENDEL

Contract Number : EE-B1-133-D

Duration : 01 January 1981 - 30 June 1983

Head of Project : M. Groll

Contractor : Institut für Kerntechnik und Energiewandlung e. V.

Address : Holderbuschweg 52, D-7000 Stuttgart 80 (Vaihingen)

Summary

Two-phase thermosyphons and gravity-assisted heat pipes are highly efficient heat transfer components for heat recovery devices. Two important aspects concerning their performance are investigated: (1) heat transfer characteristics, (2) long-term behaviour. For relatively short, small-diameter thermosyphons subject to high radial evaporator heat fluxes film boiling is the dominant heat transfer limit. For various thermosyphons the heat transfer coefficients (in the evaporator, condenser and overall) are determined. For relatively long high-performance thermosyphons subject to small radial evaporator heat fluxes vapor-liquid flow interaction is the performance-limiting effect. Flow separators are developed to suppress the flow interaction. The maximum performance of high-performance thermosyphons with and without flow separators is determined. The materials compatibility of 25 combinations of working fluids and structural materials (total of 61 thermosyphons) is investigated in automated test rigs.

1 Introduction

Gravity-assisted heat pipes and closed two-phase thermosyphons ("wickless" heat pipes) are gaining increasing importance as heat transfer components in heat recovery devices. Their heat transfer behaviour is rather complex and depending on numerous influence parameters.

For relatively short thermosyphons/heat pipes (0.5 to 1 m long) with relatively high radial evaporator heat fluxes the heat transfer is generally limited by film boiling; for small diameters ($\lesssim$ 7 mm I.D.) a dry-out limit can also occur, and vapor-liquid flow interaction plays a role. Thermosyphons/heat pipes (0.5 m long and about 10 mm I.D.) with different capillary structures and subject to relatively high radial evaporator heat fluxes are systematically investigated to determine their heat transfer coefficients. In relatively long high-performance thermosyphons/heat pipes ($\gtrsim$ 2 m) vapor-liquid flow interaction is usually limiting the heat transfer. High-performance thermosyphons/heat pipes (2.5 m long, 17 mm I.D. and 4.7 m long, 36 mm I.D.) with different surface roughnesses, with and without flow separators are investigated. Special emphasis is laid on the efficiency of flow separators.

The long-term behaviour without performance degradation requires that working fluids and structural materials are compatible. Compatibility tests are carried out, especially in the temperature range 250 °C to 400/450 °C with 25 materials combinations; a total of 61 thermosyphons is investigated in automated tests rigs. Moreover, thermosyphons from a previous CEC contract which are operating since more than two years (temperature range 80 °C to 290 °C) are included in the present tests.

2 Heat Transfer Characteristics of Investigated Heat Pipes/Thermosyphons

Heat transfer coefficients and the overall effective thermal conductance of different heat pipes and closed two-phase thermosyphons are calculated from measured thermal performance and temperature differences. The individual heat transfer coefficients for the evaporator and condenser sections, α_E and α_c, are defined as $\alpha_E = \dot{Q}/A_E \cdot \Delta T_E$ and $\alpha_c = \dot{Q}/A_c \cdot \Delta T_c$; ΔT_E and ΔT_c are the temperature drops in the evaporator and condenser sections, respectively. A characteristic figure for the thermal performance of a heat pipe/thermosyphon is its effective thermal conductance, defined as $C_{eff} = \dot{Q}/\Delta T$.

2.1 Experimental Set-Up

All investigated heat pipes/thermosyphons have a total length of 0.5 m; the lengths of the evaporator and condenser sections are 0.16 m; other relevant data are given in Table 1.

A view of the test set-up is depicted in Fig. 1. The heat input is provided by electrical heaters, viz. copper half-shells with brazed-on coax-heaters. The cooling equipment consists of a jacket calorimeter and a cryostat; the cooling medium is methanol. The heat input is measured as the product of electrical current and voltage drop at the heater. The temperatures on the outer surface of the heat pipes/thermosyphons are measured using 10 NiCr-Ni thermocouples. In some cases one additional thermocouple is installed inside the heat pipes/thermosyphons to measure the vapor temperature. All temperatures are recorded with a 12-point-recorder, and additionally the vapor temperature and the temperature at the botton evaporator wall are recorded with a line-recorder. The heat pipes/thermosyphons are well insulated to keep heat losses to the surroundings small. A detailed description of the measuring technique is given in /1/.

2.2 Results

Figs. 2 to 4 show evaporator and condenser heat transfer coefficients and the effective thermal conductance versus heat flow $\dot{Q}$, respectively, for thermosyphons AW3, BW3 and DW1 and heat pipe CW8 in nearly vertical orientation. For all thermosyphons α_E is increasing with increasing $\dot{Q}$. For CW8 α_E reaches a maximum after a sharp increase at small $\dot{Q}$. This maximum occurs at a rather small heat flow of about 250 W. Then α_E is falling with further increasing $\dot{Q}$. The maximum α_E-values for all heat pipes/thermosyphons are of the same order of magnitude. In some cases the determination of α_c was problematic because of relatively large temperature oscillations and small temperature differences in the condenser section. α_c is steadily increasing with increasing $\dot{Q}$, reaches a maximum at about 500 to 650 W and then drops off with further increasing $\dot{Q}$. Again CW8 shows a different behaviour. There α_c reaches its maximum at a rather small heat flow of about 300 W. The maximum values of α_c are about 3 times higher than the maximum values of α_E. The behaviour of the effective thermal conductance is essentially determined by α_E. Only for the highest measured heat flows α_c becomes determining. There the decreasing values of α_c lead to a decrease of C_{eff}. CW8 has a significantly higher C_{eff} up to about 400 W. This is due to the fact that AW3, BW3 and DW1 have no capillary structure or only a small surface roughness, respectively, and therefore, it is assumed that the wetting of the heated surface is better in the screen-wick heat pipe CW8.

The characteristics of the effective thermal conductances are similar for both nearly horizontal and nearly vertical orientation, i.e. for low and medium heat flows C_{eff} is determined by α_E. The thermal conductances are not very much different. At small and medium heat flows below about 350 - 400 W, CW8 has definitely the highest conductance. For higher heat flows the conductance of CW8 drops off and thermosyphons AW3, BW3, DW1 become better. For near-horizontal orientation the sequence is DW1, AW3, BW3; for near-vertical orientation the differences become minor. A summary of the experimental heat transfer data is given in Table 2. The data hold for near-optimum fill ratios $F = V_M/V_H$ (V_M = volume of liquid fill charge, V_H = inner volume of evaporator section). For most of the thermosyphons/heat pipes the temperature-dependence of the maximum radial heat flux $\dot{q}_{rad,max}$ and of C_{eff} show a similar behaviour. As to be expected, both $\dot{q}_{rad,max}$ and C_{eff} are increasing with increasing temperature. There is one exception: for CW2 the effective thermal conductance is higher at 50 °C than at 70 °C, a fact which could not be soundly explained. Both $\dot{q}_{rad,max}$ and C_{eff} are higher for near-vertical orientation than for near-horizontal orientation. The angle-dependence of $\dot{q}_{rad,max}$ is different for those thermosyphons/heat pipes, where the condensate is protected from the counter-current vapor flow and for those where the condensate is exposed to the vapor flow. In the first case $\dot{q}_{rad,max}$ is increasing from 0° to 90°, whereas in the second case $\dot{q}_{max}$ reaches a maximum at about 40° to 60°.

3 Vapor-Liquid Flow Interaction

3.1 Glass-Thermosyphon

Experimental Set-Up

The investigated glass-thermosyphon is 2 m long, with 22 mm O.D. and 18.5 mm I.D. The heat input is provided by a NiCr-ribbon heater which is wrapped around the thermosyphon along a length of 1 m. The cooler consists of a glass-jacket calorimeter, length 500 mm or 800 mm, respectively; the cooling liquid is methanol. Ten NiCr-Ni thermocouples are glued on the outer thermosyphon surface. Two additional thermocouples are measuring the vapor temperature and the liquid pool temperature, respectively. All temperatures are recorded with a 12-point-recorder; in addition, 6 temperatures are recorded with a line-recorder to pick up fast temperature changes. Fig. 5 shows a view of the test rig. Both water and ethanol are used as working fluids.

Results

With smooth inner surface and with small applied powers up to about 100 W only evaporation at the liquid surface occurs. The vapor condenses in the lower part of the condenser, the condensate returns in form of small rivulets and partly evaporates passing the upper part of the heating zone. Local dry-out occurs due to the incomplete wetting of the heated surface. For medium heat inputs (about 200 to 400 W) a large vapor bubble is formed at the bottom evaporator end and a liquid plug is expelled rather rapidly towards the condenser end. A certain amount of the condensate is collected at the top condenser end, the rest flows down towards the evaporator in a wavy flow pattern and hits the upgoing liquid plug. The periodic operation becomes weaker and finally stops. After a certain time period, e.g. about 1 min., it restarts. For higher powers (about 400 to 600 W) vapor bubbles are formed at some nucleation sites within the liquid pool. At the upper section of the pool the forming liquid-vapor mixture is periodically expelled towards the condenser. When it hits the liquid plug in the cooling section a loud noise arises due to the direct condensation of vapor bubbles in the cool liquid. For high powers (> 700 W) the vapor bubbles are generated more rapidly and violently, and a situation arises where the upgoing liquid-vapor mixture meets the downcoming condensate flow at a location shortly below the bottom condenser end. Then the upper part of the evaporator is no longer sufficiently cooled, dry-out occurs and the maximum performance is reached. With ethanol the oscillations are much weaker than with water.

Figs. 6 and 7 show the maximum performance versus operating temperature (i.e. temperature in the adiabatic section) with different fill ratios F = 0.5 to 0.7 for an inclination angle of 2° and 5°, respectively. Water at 2° has the highest performance. For ethanol the maximum performance is not much influenced by the different fill ratios. For both inclination angles the larger condenser section results in an improvement of the maximum performance of about 5 to 15 %. For the greater inclination angle the maximum performance is markedly increased due to the improved wetting of the evaporator surface. Performance limits for higher inclination angles could not be obtained as the cooling system was not able to keep the thermosyphon temperature small enough so that the system pressure did not exceed the admissible range.

In a second test series a spiral wire in good contact with the wall was installed in the heating zone to improve the evaporating heat transfer. Now the thermosyphon showed a very different flow behaviour and temperature distribution except for low heat inputs (about 100 W). For medium to high powers (200 - 600 W) many nucleation sites are activated and nucleate boiling takes place forming a two-phase mixture with fine dispersed vapor bubbles in the upper region of the pool, see Fig. 8. Weak flow oscillations still appear. The condensate returns along the wedges of the spiral wire and provides for good wetting of the heated wall above the pool. Temperature oscillations have practically ceased and all temperatures are close together; this is a desired operating state of closed two-phase thermosyphons. The liquid plug in the condenser is smaller than without capillary structure. For high powers (> 900 W) most of the condensate returns to the evaporator in a very irregular film. For horizontal and nearly horizontal orientations, local superheating of the upper part of the evaporator is diminished due to the improved wetting all around the heated surface. As the number of nucleation sites is markedly increased nucleate boiling prevails.

As the installed spiral wire in the heating section leads to diminished flow oscillations and smaller temperature differences the overall heat transfer in the glass-thermosyphon is improved.

3.2 Vapor-Liquid Flow Separators

Experimental Set-Up

Three flow separators are tested with a 2.5 m long, 20 mm O.D., 17 mm I.D. copper-water thermosyphon (EW1) with fine triangular circumferential grooves (about 0.9 mm deep, 60° opening angle, 1.7 mm pitch). One separator is a perforated copper strip, 1 mm thick, which is installed in the lower half of the thermosyphon (EW2). The second separator consists of a filled screen-wick artery (elliptic shape with 9 mm x 3 mm O.D., 100 mesh) which is installed at the bottom of the thermosyphon (EW3). The third separator is a perforated copper tube (14 mm O.D. x 0.65 mm thick) which is concentrically installed in the thermosyphon (EW4). There are 4.5 mm holes (pitch 6.4 mm) in the evaporator section and 3.5 mm holes (pitch 5.5 mm) in the condenser section; there are no holes in the 0.5 m long adiabatic section. Fig. 9 shows the cross sections of thermosyphons EW1 to EW4. Heat input to the thermosyphon is provided by a coaxial heater wire (3 mm O.D.) which is brazed onto the 1 m long evaporator section. Two annulus calorimeters, 0.5 m long each, are used. Water which is cooled by a methanol cryostat is the coolant. The coolant is in direct contact with the condenser surface. Twelve temperatures are measured with Ni-CrNi thermocouples and recorded by a 12-point-recorder. A 10 mm thick asbestos-foam insulation keeps thermal losses to a minimum. A view of the test rig is shown in Fig. 10 (left side).

Results

In Figs. 11 and 12 the performance of the various thermosyphons with flow separators is compared with that of a thermosyphon without flow separator (EW1). The perforated metal strip separator (EW2) results in a poor performance compared to EW1. EW2 is only useful for $\beta < 2°$. The eccentrically located screen-wick separator (EW3) shows a better performance for $\beta \gtrsim 20°$; its performance for $\beta \lesssim 20°$, however, is poor. For all inclination angles the performance of EW3 is below that of EW1. The concentrically located perforated tube separator (EW4) results in an excellent performance for $\beta \gtrsim 1°$. For $\beta \gtrsim 2°$, its performance is already significantly better than that of EW1. A maximum performance of nearly 4 kW (at 60 °C) is reached for $\beta \gtrsim 45°$. This performance remains about constant till $\beta = 90°$.

3.3 High-Performance Thermosyphon

Experimental Set-Up

A closed two-phase copper-water thermosyphon (4.7 m long, 40 mm O.D., 36 mm I.D.) with technically smooth inner surface (GW1) was tested as reference case for a planned test series comprising roughended surfaces and a concentric perforated tube flow separator. The heating system consists of six pairs of copper half-shells, each 0.34 m long, with brazed-on coax heater wires. The condenser section is cooled by 2 annulus calorimeters, each 1 m long. Water as the coolant fluid is in direct contact with the condenser surface. The adiabatic section is about 0.5 m long. The axial temperature distribution is measured with 12 Ni-CrNi thermocouples using a 12-point-recorder. A view of the test set-up is shown in Fig. 10 (right side).

Results

The maximum performance of GW1 was measured for various fill charges. The obtained experimental data were extrapolated to higher temperatures using the figure-of-merit for thermosyphons /2/. Measured and extrapolated results are shown in Fig. 13. Best results are obtained for an optimum fill ratio of about F = 0.7 to 0.75.

4 Materials Compatibility

For the materials compatibility tests a temperature range of about 250 °C to 400/450 °C was the major objective. Heat pipes from a previous CEC-contract (no. 379-77-10-EED) which are operating since more than 2 years in the temperature range of 80 °C to 290 °C have been included in the present tests.

4.1 Experimental Set-Up and Data Assessment

In Table 3 the investigated working fluid/structural material combinations are listed together with first test results. Most of the new 61 heat pipes are installed in the two new automated tests rigs (Fig. 14) and are now in operation.

During the tests the temperature distribution of the heat pipes is measured with Ni-CrNi-thermocouples. Fig. 15 shows the instrumentation. A cold end of the heat pipe or a superheating of the evaporator section indicate a degradation or failure; the cause can be noncondensable gas generation due to thermal decompostion of the working fluid or chemical or electrochemical reactions, deposits in the evaporator due to impurities or corrosion products. Continuous measurement of the heat pipe temperatures especially of the temperature difference between evaporator and condenser is a reliable method to check such potential degradation effects.

The electronic data unit coupling the test rigs to the laboratory computer has been built and different user programmes for assessing, listing and plotting the measured temperatures have been generated. A schematic of the electronic data unit is shown in Fig. 16. This unit contains a CPU based on a Z80 microprocessor as well as data (RAM) and programme (ROM) storage. The voltages of the thermocouples (six per heat pipe) are measured via a multiplexer, first amplified by a pre-amplifier and then, according to the measurement range, by a programmable gain amplifier. After the measurements have been converted to a digital output, the data are stored in the RAM. The whole measurement cycle is started and controlled by the laboratory computer PDP 11/40.

4.2 Results

Table 3 shows the temperature differences T_1-T_5 and T_1-T_6. The following materials combinations, operating now since about 4 years, have been found to be perfectly compatible: freon/CuNi30Fe, freon/CuZn36, freon/CuZn20Al, freon/St35; water/CuNi30Fe; n-octane/St35, and n-octane/X10CrNiTi189. The other materials combinations (freon/AlMn, freon/AlMgSi0.5 freon/CuZn28; water/SF-Cu; diphyl/St35 and diphyl/X10CrNiTi189) which are also operating since about 4 years developed a temperature difference of $\gtrsim$ 10K. Only the combination water/CuZn20Al is absolutely incompatible (temperature difference > 100K). The heat pipe n-octane/St35 which is operating at 170 °C is sealed by a stainless steel valve. An electrochemical potential difference exists between this valve and the mild steel container. Thus a corrosion reaction is caused which is responsible for the great temperature differences T_1-T_5 and T_1-T_6. The heat pipes toluene/St35, 13CrMo44 and X2CrNiMo1812 which are operating now for about 2 years show a good compatibility. Naphthalene seems to be compatible with all tested steels. Silicon oil 200 fluid 50 CS is not compatible together with St35, 13CrMo44 and CuNi10Fe. In case of the other heat pipes a clear statement cannot yet be given because the operating duration is too short.

5 Summary and Conclusions

The thermal behaviour of gravity-assisted heat pipes and closed two-phase thermosyphons is still not perfectly understood and no reliable design tools are at hand. Detailed quantitative information on both maximum performance and heat transfer coefficients is still lacking. This information, however, is of utmost importance for an optimum and cost-efficient design of heat recovery devices employing such heat transfer components. Concerning the long-term behaviour of heat recovery devices, the compatibility of working fluids and structural materials and the thermal stability of the working fluids are decisive. In the temperature range > 250 °C to about 450 °C, which becomes more and more important for industrial heat recovery, no suitable working fluids (except mercury) are available.

The work carried out is an attempt to fill the above-described information gaps. For relatively short small-diameter thermosyphons/heat pipes with relatively high radial evaporator heat fluxes quantitative heat transfer data are compiled and design tools generated. First results on the efficiency of flow separators in relatively long high-performance thermosyphons/heat pipes with relatively small radial evaporator heat fluxes are obtained, as well as a better understanding of the vapor-liquid flow interaction. However, in this field there remains much to be done till the state-of-knowledge becomes satisfactory.

Materials compatibility tests provide long-term information (more than 2 years) on various suitable combinations of working fluids and structural materials for temperatures below about 250 °C. A new test series for higher temperatures shall provide respective information on potentially suitable materials for the temperature range of about 250 °C to 450 °C.

References

/1/ M. Groll, H. Nguyen-Chi, H. Krähling: Wärmerückgewinnungsanlagen mit Reflux-Wärmerohren als Bauelementen. Schlußbericht Vertrag Nr. 379-77-10 EED, Bericht Nr. IKE-5TF-363-80, März 1980

/2/ ESDU Item No. 80017: Thermophysical properties of heat pipe working fluids: operating temperature range between -60 °C and 300 °C. London, 1980

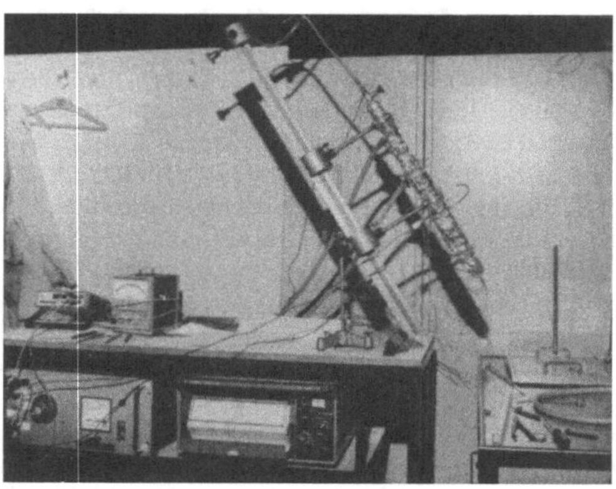

Fig. 1: View of test set-up:
short thermosyphons (AW,BW,CW,DW)

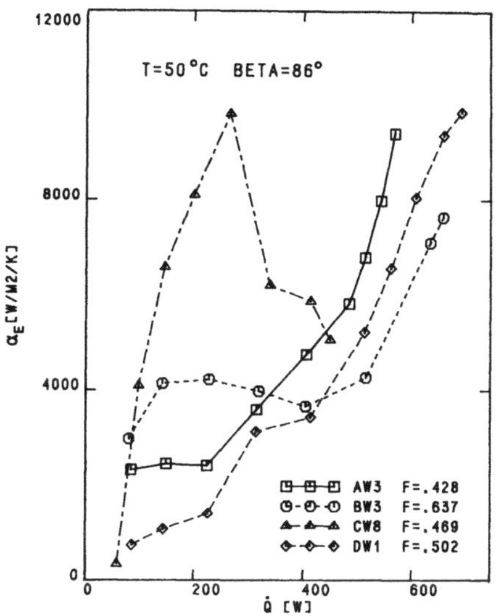

Fig. 2: Evaporator heat transfer coefficients

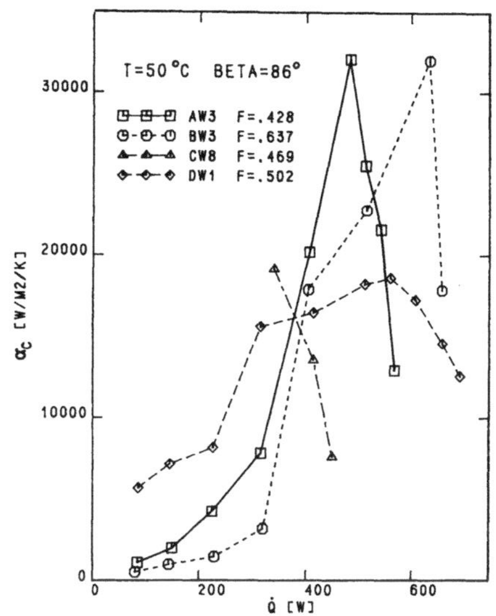

Fig. 3: Condenser heat transfer coefficients

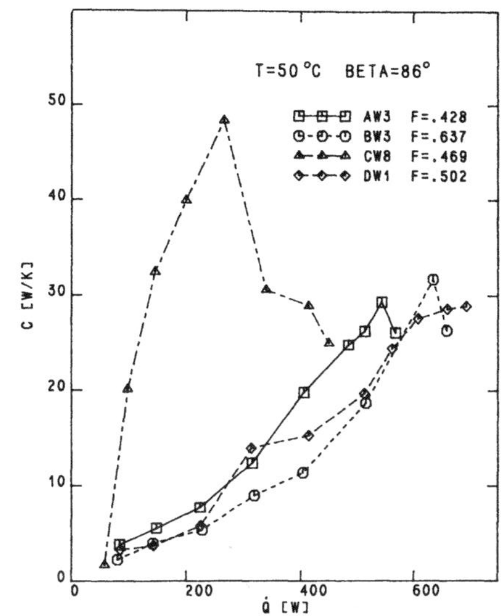

Fig. 4: Effective thermal conductance

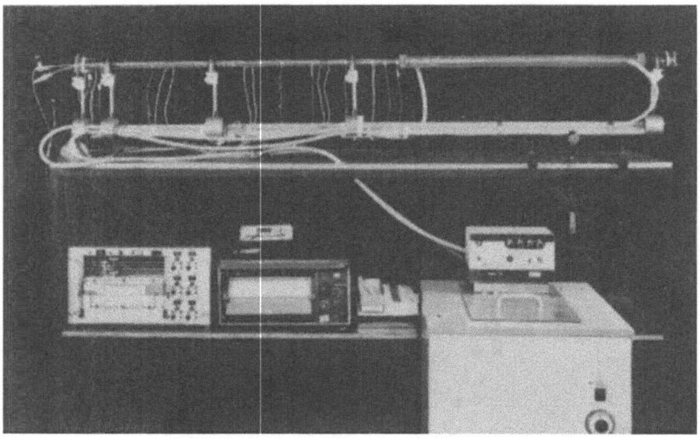

Fig. 5:
View of test set-up:
glass thermosyphon

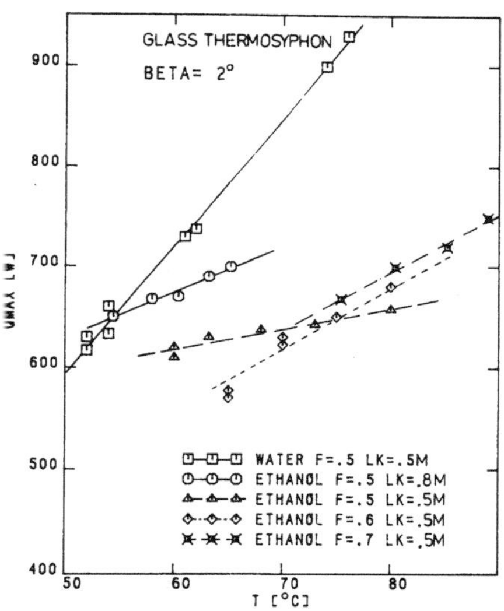

Fig. 6: Maximum performance of glass thermosyphon

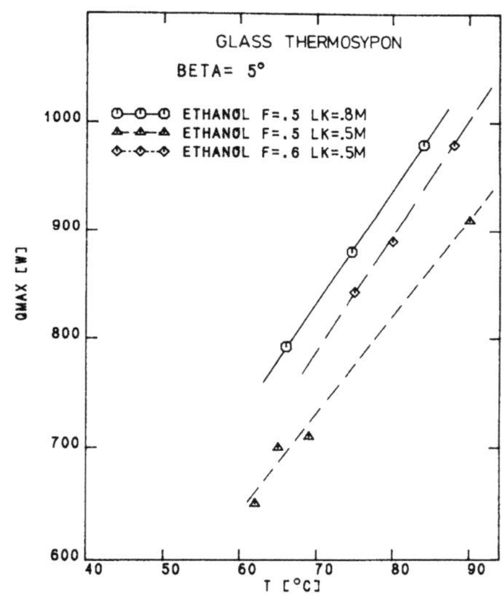

Fig. 7: Maximum performance of glass thermosyphon

Fig. 8: Nucleate boiling in evaporator of glass thermosyphon

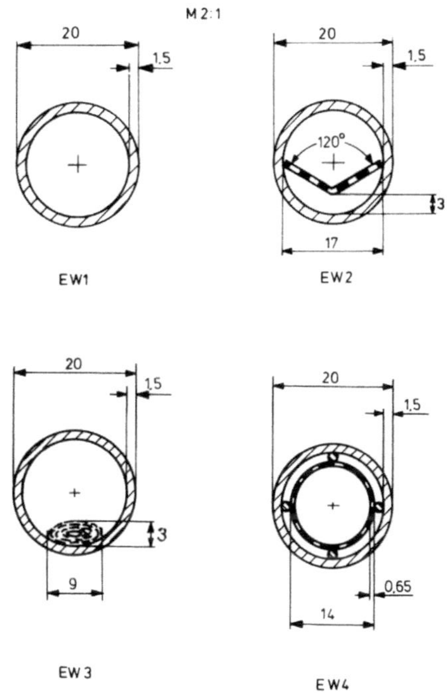

Fig. 9: Cross sections of thermosyphons EW1, EW2, EW3, EW4

Fig. 10: View of test rigs.
Left: 2.5 m-thermosyphon (EW),
right: 4.7 m-thermosyphon (GW)

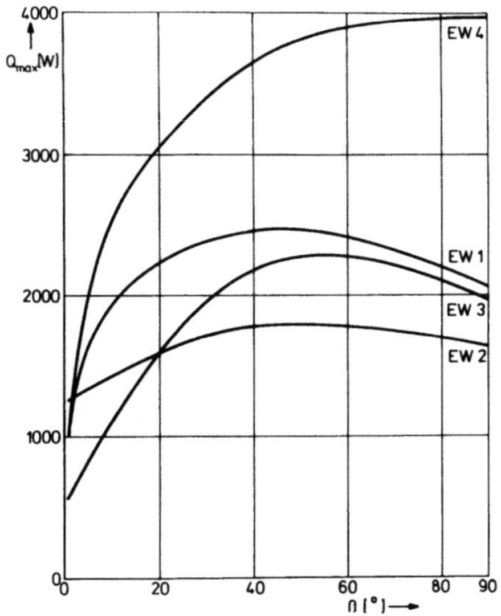

Fig. 11: Maximum performance of thermosyphons EW1, EW2, EW3, EW4

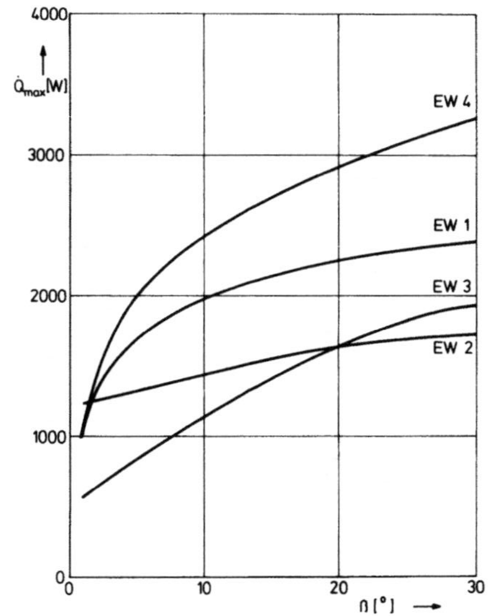

Fig. 12: Maximum performance of thermosyphons EW1, EW2, EW3, EW4

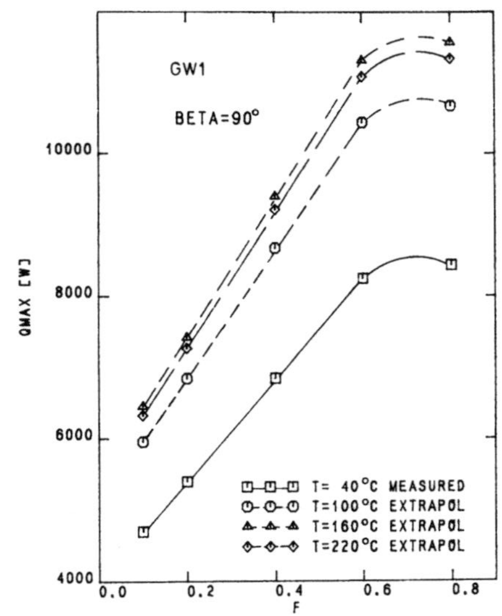

Fig. 13: Maximum performance of thermosyphon GW1

Fig. 14: View of compatibility test rigs

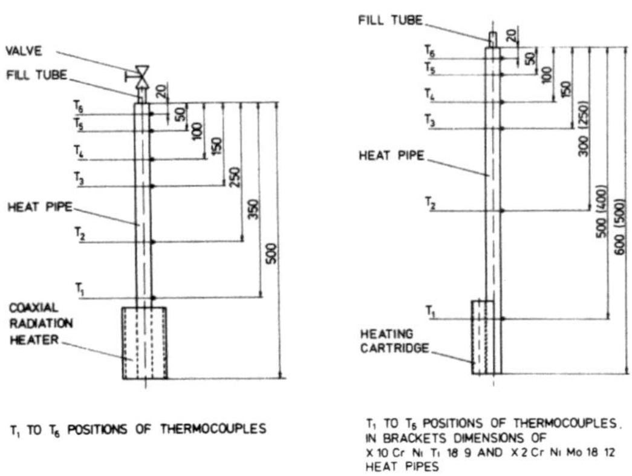

a) for T $\leq$ 150 °C   b) for T > 150 °C

Fig. 15: Instrumentation of compatibility test heat pipes

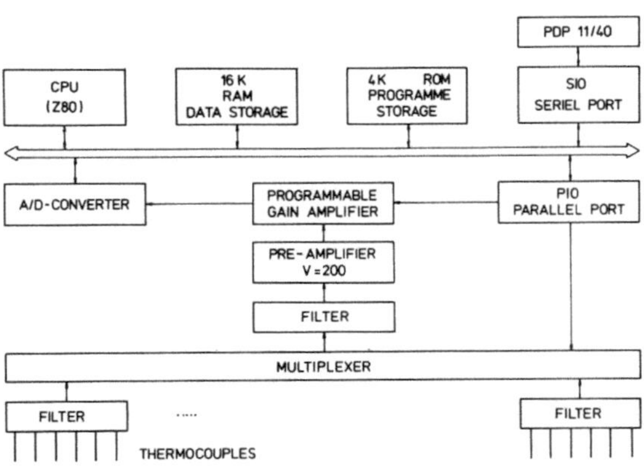

Fig. 16: Schematic of electronic data unit

Tab. 1: Data of the Investigated Heat Pipes/Thermosyphons

Heat Pipe/ Thermosyphon	Inner Diameter /mm/	Outer Diameter /mm/	Capillary Structure
AW1	7.0	9.0	no capillary structure
AW3	10.0	12.0	" " "
AW4	5.0	7.0	" " "
AW5	11.5	13.5	" " "
			Chemically roughened
BW1	10.0	12.0	roughness 6.0 µm
BW2	10.0	12.0	" 7.0 µm
BW3	10.0	12.0	" 14.5 µm
BW4	10.0	12.0	" 17.0 µm
CW2	11.5	13.5	2 layers of 160 mesh screen
CW3	11.5	13.5	4 " " 160 " "
CW4	11.5	13.5	6 " " 160 " "
CW5	11.5	13.5	2 " " 50 " "
CW7	13.5	15.5	2 " " 160 " "
CW8	9.8	11.8	2 " " 160 " "
DW1	10.0	12.0	65 axial grooves, 75 µm deep triangular shape
DW2	10.0	12.0	circum ferential grooves, 75 µm deep, triangular shape

TAB. 2: OVERVIEW ON HEAT TRANSFER DATA FOR VARIOUS HEAT PIPES AND THERMOSYPHONS

HEAT PIPE / THERMOSYPHON (FILL RATIO F)	MAX. RADIAL HEAT FLUX $\dot{Q}_{RAD,MAX}$ $(W/M^2)x10^{-4}$			EFF. THERMAL CONDUCTANCE $C_{EFF}(W/K)$ AT $0.9X\dot{Q}_{RAD,MAX}$			TEMP. (^oC)
	$\beta = 10^0$	$\beta = 45^0/50^0$	$\beta = 86^0$	$\beta = 10^0$	$\beta = 45^0/50^0$	$\beta = 86^0$	
AW1 (0.856)	2.9	4.5	6	3.0	3.7	5.5	20
	5.2	9.1	8.7	5.5	5.4	7.6	50
AW4 (1.116)	0.8	3	3.2	1	2.2	1	30
	1.5	4.2	5.3	2.2	2.8	3.5	50
	3.5	6.7		3.6	4.9		70
AW5 (0.50)	6.3		10.8	10.3		29.5	30
	8.5		13.4	34		44	50
	10.9		17.2	29		64	70
BW1 (0.597)	5		9.2	14.4		20.9	30
	7.6		12.6	32.6		30	50
	8.6		14.8	35.2		37.5	70
BW2 (0.422)	5		9	15.6		31.3	30
	6.9		13	24		26.7	50
	9.8		17.2	38		48	70
BW4 (0.454)	5.7		10.3	8.4		14.1	30
	8.4		12.9	14		17.4	50
	10.1		15.2	23.4		19.2	70
CW2 (0.209)	5.3	7.5		22.7	41		30
	5.9	8.3	10.5	38.1		85	50
		10.4	12.9		83	75	70
CW3 (0.165)	4.5			28.2			30
	5.5			45.8			50
			11.6			64	70
CW4 (0.25)	4.1		6.2	26		33	30
			8.8			62	50
	6.2			36			70
CW5 (0.285)	5.5		9	52		50	30
CW7 (0.341)	7.2		10.1	23.1		35	30
	10.3		13.5	41		48	50
	13.6		15.6	68		55	70
DW1 (0.502)	4.4		9.4	6.8		14	30
	10		15.9	43		25.7	70
DW2 (0.576)	4.9		8.5	14.3		19.2	30
	9		12.7	24.3		32.8	50
			15.1			34.2	70

TAB. 3: RESULTS OF COMPATIBILITY TESTS

Working Fluid		Structural Material			Operating Temperature	Duration[1] /d/	Temperature Differences	
Name	Chemical Formula	DIN	German material-number	AISI-number			$T_1 - T_5$ /K/	$T_1 - T_6$ /K/
Freon R 21	$CHCl_2F$	AlMn	3.0515		80	1430	10-23	16-23
		AlMgSi0.5	3.3206		80	1430	25-30	25-36
		CuNi30Fe	2.0882		80	1430	0-2	0-2
		CuZn28Sn	2.0470		80	1430	2-11	5-16
		CuZn36	2.0335		80	1430	2-3	5-11
		CuZn20Al	2.0460		80	1430	0	0-2
		St35	1.0305		80	1430	0	0
Water	H_2O	SF-Cu	2.0090		150	1450	9-11	8-16
		CuNi30Fe	2.0882		140	1440	0	0
		CuZn20Al	2.0460		150	1440	101-114	104-126
		CuNi10Fe	2.0872		300	30	0-1	0-1
		Ti994	3.7025		300	30	0-1	0-2
n-Octane	C_8H_{18}	St35	1.0305		170	1350	143	144
					230	1350	0	0
		X10CrNiTi189	1.4541	321	200	1350	3	4
					250	1350	0	0
Diphyl[2]	$C_{12}H_{10}$ + $C_{12}H_{10}O$	St 35	1.0305		250	1350	12	14
					270	1350	12	37
		X10CrNiTi189	1.4541	321	270	1350	24	23
					290	1350	28	56
		CuNi10Fe	2.0872		300	30	4-8	4-8
		Ti994	3.7025		300	30	3-5	14-17
Diphyl-O[3]	$C_6H_4Cl_2$	St35	1.0305		220	1360	6-13	6-14
Diphenyl	$C_{12}H_{10}$	13CrMo44	1.7335		350	10	2-3	3-6
					400	10	1-2	1-2
		X2CrNiMo1812	1.4435	316L	350	10	1	1
					400	10	4	7
OM[4]		13CrMo44	1.7335		350	*	-	-
					400	*	-	-
		X2CrNiMo1812	1.4435	316L	350	*	-	-
					400	*	-	-
OMD[5]		13CrMo44	1.7335		350	*	-	-
					400	*	-	-
		X2CrNiMo1812	1.4435	316L	350	*	-	-
					400	*	-	-
Toluene	C_7H_8	St35	1.0305		250	730	0	3-5
		13CrMo44	1.7335		250	730	7	1-12
		X2CrNiMo1812	1.4435	316L	280	600	0-3	0-2
		CuNi10Fe	2.0872		300	90	1-3	2-3
		Ti994	3.7025		300	30	1-3	2-4
Naphtalene	$C_{10}H_8$	St35	1.0305		270	730	1-4	3-7
		13CrMo44	1.7335		270	730	3-7	8-13
		X2CrNiMo1812	1.4435	316L	350	20	4-5	1-6
		CuNi10Fe	2.0872		350	30	0-5	1-6
		Ti994	3.7025		300	30	3	3-4
Silicon Oil XF 218[6]		St35	1.0305		300	80	7-10	12-14
		13CrMo44	1.7335		350	80	9-14	10-18
		CuNi10Fe	2.0872		350	80	11	15
Silicon Oil 200 Fluid 50 cs[6]		St35	1.0305		300	80	90-104	113-122
		13CrMo44	1.7335		350	80	143-108	163-181
		CuNi10Fe	2.0872		350	80	104-109	123-131
Sulphur + Iodine 99 % + 1 % 97 % + 3 %	S (99 %) + I (1 %) S (97 %) + I (3 %)	AlMg 3	3.3535		200	*	-	-
		AlMgSi0.5	3.3206		200	*	-	-
		X2CrNiMo1812	1.4435	316L	450	*	-	-
					500	*	-	-
					550	*	-	-

1) heat pipes: * not in operation
2) binary mixture of diphenyl and diphenyl oxide
3) 1,2-dichlor benzene
4) binary mixture of ortho- and meta - terphenyl
5) ternary eutectic mixture of diphenyl, ortho- and meta- terphenyl
6) trade name of Dow Corning Company

DISCUSSION

There is considerable interest in high temperature organic working fluids, and this was evident in the discussion.

Question - Mr. Galant, Bertin : Concerning your work on toluene, could you elaborate on experimental work above 300°C, especially related to fluid stability ?

Answer So far our experiments with toluene have been carried out at temperatures no higher than 300°C. No fluid stability problems occurred for test durations of 2 years at 280°C and 3 months at 300°C.

Question - Mr. Galant, Bertin : Could you define fluid-metal compatibility? (does it mean fluid stability and/or metal stability?) Do you take account of possible catalytic effects on decomposition rates ? Do you take account of small impurities within the fluid which are inherent in the fabrication process ?

Answer Fluid-metal compatibility basically comprises both aspects - fluid stability and reactions between fluid and metal, i.e. corrosion. Possible catalytic effects are accounted for insofar as the heat pipes are provided with fine circumferential grooves, resulting in a large surface area. The fluid is inserted into the heat pipes as obtained from the manufacturer, without applying extra purification procedures, e.g. distillation. The manufacturer's specification for Toluene is :

C_7H_8	:	99.7%
C_4H_4	:	0.0005%
H_2O	:	0.02%
non-volatiles	:	0.0005%

Before filling the heat pipes, the toluene is exposed to air for some minutes. The heat pipes themselves are degreased using 111-trichloroethane.

Question - Dr. Pilavachi, CEC : Have you experience of, or do you know, fluids which could be used in heat pipes at temperatures above 400°C.

Answer 1. Various organic fluid heat pipes have been developed at IKE. An extended organic fluid compatibility test programme is in progress. The test durations however, are in general too short to make conclusive statements.

According to information from fluid manufacturers the thermal stability of Diphenyl is about 400-420°C. We have, therefore, reduced the test temperatures for our organic fluid heat pipes to <420°C. In our tests with Diphenyl heat pipes, operating at 400°C, we observed serious fluid decomposition after a short accidental overheating of the heat pipes (maximum temperatures of 420-450°C may have been reached for about 0.5 - 1 hour).

2. Mercury is a useful working fluid with excellent thermo-physical properties. However, its toxicity, cost, and availability may be prohibitive for its general application. Mercury heat pipes have not been developed at IKE.

3. Alkali metals (K, Na) are standard working fluids for temperatures above about 500 to 550°C. Their heat transport capability is relatively small below about 600°C (vapour pressures of 180 mbar for K and 33 mbar for Na). IKE has a long experience in alkali metal heat pipe technology.

DESIGN, BUILDING AND TESTING OF A NOVEL MODULAR GAS-GAS CERAMIC HEAT EXCHANGER FOR EXHAUST HEAT RECOVERY AROUND 1200°C

Authors	: D. GROUSET, G. MARTINEZ, J.C. MULET, D. REBUFFAT, S. GALANT
Contract	: EER-1-107-F
Duration	: 18 months : 1 March 1981 - 30 August 1982
Head of project	: S. Galant, Division Energétique
Contractor	: Société Bertin & Cie
Address	: 78370 PLAISIR, France

Summary

In many industrial furnaces, one way of obtaining energy savings is the recovery of waste heat by using recuperators to preheat the combustion air to the furnace burners. An added advantage of preheated air is improved efficiency of the combustion process. A novel gas to gas ceramic heat exchanger developed by Société BERTIN is presented. The following goals are reached : capability to handle exhaust gases at temperatures up to 1200°C, large overall heat transfer coefficients (viz. 150 $W/m^2 °C$ and above) by using both radiative and jet impingement convective heat transfer, capability of being assembled in the form of modules, prospective cost which must be comparable to metallic heat exchangers currently available at much lower temperatures (800°C). A thermal equilibrium mathematical model of the heat exchanger module is set up to optimize the module efficiency and pressure drops as a function of jet impingement heat transfer characteristics, hot and cold gas flow characteristics, ceramic material (silicium carbide) properties. A thermo-mechanical study conducted with CERAVER (France) leads to the following optimal design for SiC made core (0.340 m long, 0.250 m wide, 0.080 m high): module efficiency 0.2 to 0.3, module NTU : 0.2 to 0.3, overall module heat transfer coefficient around 150 $W/m^2 °C$, module pressure drop (air side) : 200 to 300 Pascal. Further testing of the core is needed to conclude about the proposed design both on technical and prospective economical grounds.

1. INTRODUCTION

In many industrial furnaces, the furnace exhaust gases are directly vented to the stack : furnaces that run at temperatures of about 1000°C or above lose about 50 % of the input energy in the flue gas exhaust, or more.

One way of obtaining energy savings is the recovery of waste heat by using recuperators to preheat the combustion air to the furnace burners. An added advantage of preheated air is improved efficiency of the combustion process.

Société BERTIN is developing a novel gas-gas ceramic heat exchanger with the following characteristics :

- capability of handling exhaust gases at temperatures up to 1200°C, and future goals up to 1500°C,
- large overall heat transfer coefficients by using both radiative and jet impingement convective heat transfer,
- capability of being assembled in the form of modules, to operate over the wider possible range of industrial applications,
- cost objectives comparable to metallic heat exchangers currently available at much lower temperatures (800°C).

Ceramic materials were screened to finally end up with Silicium Carbide, a CERAVER made material. In the present progress report, a summary of the methodology (viz. computational efforts as well as experimental program) and preliminary results are presented, together with the final experimental program and economical assessment which are to be carried out before the end of the program.

2. PROGRAM OBJECTIVES (1, 2)

The program proposed by BERTIN is divided into 4 complementary sub tasks.

Sub task 1 : a review of existing industrial needs is performed to asses :

- potential applications in various processes whose exhaust gas temperatures lie above 1000°C
- load spectra for those furnaces including temperature and flow rate data, hot gas composition and particulate concentrations
- present day technology (viz. metallic heat exchanger) with price quotes to set economical constraints on the BERTIN project.

Sub task 2 : system analysis and thermal optimization of the heat exchanger configurations which goes through :

- the building of a mathematical model which describes local as well as global energy balances
- experimental "cold" tests (viz. below 600°C) which allows comparisons with theoretical results as found by the mathematical model as well as further refinements to adequately predict overall heat exchange data (± 15 % accuracy)
- a step-by-step optimization study to end up with various designs to be checked in the context of the mechanical study.

Sub task 3 : thermo-mechanical study jointly with CERAVER to end up with a first design to be tested on a thermal test rig. The following steps are considered :

- size study to yield overall dimensions that are compatible with existing cold pressing equipment at CERAVER

- optimization of ceramic plate fabrication processes
- building of the heat exchanger prototype

Sub task 4 : experimental program which considers :

- check of mechanical integrity
- steady state efficiency measurements as a function of temperatures differences between hot and cold stream

Finally, conclusions are to be drawn on the proposed design both on technical and prospective economical grounds.

3. RESULTS OBTAINED AS OF JUNE 1st 1982

In the following, a summary of the results obtained for completed sub-task 1 to 3 is presented. Sub-task 4 will end up on August 30th 1982 : the final report dealing with the experimental results will then be delivered to EEC program management

3.1. Survey of industrial requirements

3.1.1. Present needs and technology

The survey dwells upon:

- furnaces whose exhaust gases lie above 1000° C with no heat recovery
- potential in situ use of extracted heat as hot air to the burners

The table below lists various industrial cases as a function of exit temperature and thermal output power (*** important, ** average, * low).

PROCESS	EXHAUST GAS TEMPERATURE °C	POTENTIAL RECOVERY
Glass melting	1400	***
Founderies	1500	***
Reheating furnaces	1000 to 1400	***
Hydrocarbons reforming	900	**
Ceramic furnaces	1000 to 1400	**
Zinc and Titanium	900 to 1500	*
Refractory oxide reduction furnaces	1200	*
Nonferrous second fusion furnaces	800 to 1000	*,**

Furthermore a special attention is paid to heat recovery on reheating furnaces, where metallic heat exchangers are used with air diluted exhaust gases (800°C). The following conclusions are drawn :

- present day metallic heat exchangers have efficiencies around 0.5. It appears realistic to reach 0.6 to 0.7 with the BERTIN concept
- present day metallic heat exchangers are optimal on an economical standpoint ; for a given installation cost, a trade-

off between thermal balances and operating costs yield minimum payout time at efficiencies around 0.5

- investment costs (including installation) should not exceed 1700 French Francs/kWth recovered, at nominal power between 1 and 10 MWth and maximum temperatures of 1500°C. For a comparison, recent data on cordierite made core(3) exhibits investment costs around 1200 French Francs at nominal power between 100 and 200 kWth and maximum temperatures of 1000°C.

3.1.2. Operating constraints

Four constraints should be considered in order not to impair the industrial usefulness of the heat exchanger :

i) overall pressure drops on the air side should not exceed 5000 Pascal. On the exhaust side, hot blowers are used to exhaust combustion gases adequately

ii) thermal cycling frequency over the heat exchanger lifetime : average load factors peak around 60 percent. Yet it is necessary to oversize heat exchanger capacity in order to have to whole system suited for 100 % percent load factor and/or accidental operating circumstances

iii) particulates are present in hot gas streams especially when heavy residual fuel oils are burnt. Particulate concentrations may reach 25 mg/MJ thermal with average sizes between 10 and 50µm. Preliminary assessment of the novel heat exchanger design should be carried out in the light of recent BERTIN work on heat exchanger fouling (4).

iv) gas and solid phase corrosion phenomena on the hot side should be carefully considered. Sulfur corrosion is of paramount importance when coke oven gas and residual oils are burnt. Extensive testing of SiC corrosion resistance shows that no serious weakness is to be expected. Yet particulate corrosion (via vanadium, nickel or borosilicates) have to be studied within the heat exchanger development program (in situ tests) (5).

3.2. System analysis and thermal optimization

The heat exchanger design is based upon four principles :

- the use of radiative heat exchange at elevated temperatures (recall that black body equivalent heat transfer coefficient at 1000°C is near 500 W/m^2°C)
- the improvement of convective heat exchange via impingement heat transfer (in Ref (6) it is shown that heat transfer coefficients much above 300 W/m^2°C can be obtained within reasonable pressure drops)
- the coupling of both radiative and convective heat exchange in order to reach elevated overall heat exchange coefficients with no preferential driving forces
- the use of elevated heat transfer coefficients to minimize heat transfer areas, hence mechanical problems related to ceramic material use.

3.2.1. Local optimization

These basic principles lead to the design sketched on Fig. 1. A series of ceramic perforated screens are positioned above a plain ceramic heat transfer surface. Air is passing through these screens which are therefore cooled down : past the last screen, air is impinging upon the final ceramic plate. A coupled radiative and convective thermal equilibrium is obtained which results in varying thermal efficiencies (for a module of length ℓ) with varying pressure drops.

In Ref (6) a local mathematical model is described which takes account of mass, momentum and energy balances for an infinitesimal slice of gas-gas heat exchanger. Only the main flow direction is considered, i.e. no two-dimensional flow characteristics are explicitely taken into account. BERTIN impingement heat transfer correlations are used throughout (6). The model allows for a local knowledge of the gas temperature, velocity and pressure fields and the local temperature field along the heat exchange plate.

This mathematical model was compared with "cold" experiments carried out on a low temperature metallic heat exchanger module (see Fig. 2) : as many as 3 metallic perforated screens are positioned above a heat exchange plate which is electrically heated. Air is injected to cool down the whole system, whereas thermal balances are made for each tested configurations. 27 configurations have been studied (1) covering various plate open areas, plate numbers, plate to plate distances, cold air temperature and air flow rates. On Fig. 3 is shown a comparison between experimentally obtained overall heat transfer coefficients and calculated ones as a function of average surfacic mass flow rates. Model and experimented results are within an average ± 15 % error estimate which is adequate for proper industrial scaling up.

3.2.2. Overall optimization

Once the local model is thorougly tested against a set of experimental results, it can be included in a more general computer program which takes account of heat exchange on both the cold air and hot exhaust sides.

For one side the independent geometric parameters are :

- the number of plates k
- the jet plate to plate distance Z_k
- the slot or jet width B_k
- the slot center to center spacing X_k
- the frontal area S
- the discharge slot coefficient C_{dk}
- the plate length ℓ

The independent flow parameters are :

- the meanflow static pressure P
- the mass flow rate Q_m
- the mean flow temperature T

The computer program can then be used to calculate for the air or exhaust side :

- the friction factor $f = \dfrac{2\rho\Delta p}{\left(\dfrac{Q_m}{S}\right)^2} (s\, C_d)^3$ (ρ: gas density)
- the Colburn modulus $j = \dfrac{h\, s\, C_d}{\dfrac{Q_m\ C_p}{S}} P_r^{2/3}$
- the length parameter $L = r_h \left(\dfrac{1}{j}\right)\left(\dfrac{h\,S}{Q_m}\right)\dfrac{P_r^{2/3}}{C_p}$

- the area parameter $A_{pa} = \left(\frac{1}{\sqrt{2}}\right)\left(\frac{1}{\sigma}\right)\left(\frac{f}{j}\right)^{0.5}\left(\frac{Q_m\ hS}{\rho\ .\ \Delta p}\right)^{0.5}\frac{Pr^{1/3}}{C_p{}^{1/2}}$

- the volume parameter $V = L\ A_{pa}$

They are a function of the following dependent variables :

h	$= f\ (X/B,\ Z/B,\ Re_{B_1})\ Pr$	heat exchange coefficient
r_h	$= s\ Cd\ Z$	hydraulic radius
σ	$= s\ Cd$	area ratio
Re	$= \frac{4\ Q_m\ Z}{S\ \mu}$	Reynolds number(μ : gas viscosity)

On Fig. 4 is shown a computer result in the case of the final design as described below. It is seen that the f/j ratio is quite good even though the pressure drop model tend to largely over-estimate the pressure drops through the plate network.

The present numerical model can be run at very low cost (10 seconds CPU time per configuration on a PRIME 650 computer).

3.3. Mechanical optimization

The mechanical optimization is conducted jointly with Société CERAVER (France). It is an iterative study which considers :

- the choice of the ceramic material : Silicium Carbide
- checking the system mechanical integrity both in static and transient operational modes (check of maximum stress yield and maximum thermal expansion)
- choosing proper fabrication processes with considerations on maximum allowable plate sizes (compatible with CERAVER cold pressing facilities), extrusion and/or molding possibilities, green state machining necessities
- designing a metallic housing which insures proper insulation of the hot parts, compression loading of the ceramic parts and gaseous stream connections with the hot and cold stream.

The final configuration is sketched on Fig. 5 which illustrates the internal plates and stream connection assembly. Fig. 6 is an artist view of the final module which underlines the role of the metallic housing together with the final flow arrangement. An inclined plate design is chosen for the following reasons :

i) it minimizes heat exchanger length

ii) it allows to design proper sealing systems between hot and cold gas streams with self compensating arrangements which take account of cold and hot side expansion ratios

iii) it yields a simple mechanical connection system which can take care of side-by-side arrangements of ceramic half modules.

Patents for the design are pending.

3.4. Final first generation design

Optimal design leads to the following SiC made core characte-

ristics :
- length : 0.340 m
- width : 0.250 m
- height : 0.080 m
- material thickness : around 0.003 m
- module efficiency : from 0.2 to 0.3
- module NTU : from 0.2 to 0.3
- overall module heat transfer : around 150 W/m^2°C
- module pressure drop (air side) : 200 to 300 Pascal

3.5. Final experimental test program

The module will be delivered at BERTIN testing facilities (BAYONNE) mid-June 1982. The hot tests will be carried out on an experimental test rig described on Fig. 7.

- a propane burner allows generation of hot combustion gases near 1200°C at flow rates compatible with heat transfer module design. Temperature and/or flow rates can be varied by adjusting the burner thermal output and its equivalence ratio
- an electric heater provides hot air (up to 600°C) at temperatures and flow rates compatible with the module "cold" air flow characteristics.

The test rig is instrumented to carry out standard thermal and aerodynamic balances which give full experimental data concerning the module capabilities. Furthermore, final high temperature tests of the computer program are to be carried out in order to make the computational technique fully operational on an industrial basis.

Testing will last one month. It is supposed to consider :

- the system mechanical integrity both at low and high temperatures
- steady state efficiency measurements as a function of temperature differences between hot and cold stream (compatible with the module mechanical integrity)
- conclusion on the proposed design both on technical and prospective economical grounds.

4. CONCLUSION

In many industrial furnaces, the furnace exhaust gases are directly vented to the stack : furnaces that run at temperatures of about 1000°C or above lose about 50 % of the input energy in the flue gas exhaust, or more.

One way of obtaining energy savings is the recovery of waste heat by using recuperators to preheat the combustion air to the furnace burners. An added advantage of preheated air is improved efficiency of the combustion process.

BERTIN & Cie is developing a novel gas to gas ceramic heat exchanger with the following characteristics :

- capability to handle exhaust gases at temperatures up to

1200°C

- large overall heat transfer coefficients by using both radiative and jet impingement convective heat transfer,
- capability of being assembled in the form of modules,
- cost of objectives comparable to metallic heat exchangers currently available at much lower temperatures (800°C),

The program funded in part by EEC is divided into 4 phases.The first of these is to carry out a survey of typical industrial needs with respect to heat recovery potentialities above 1000°C.

Both operational constraints (load spectra, exhaust composition including particulate concentrations) and investment cost objectives are obtained, in order to set realistic goals for the BERTIN & Cie project.

Next, a theoretical and experimental low temperature (metallic) concept study is carried out to develop a mathematical model describing thermal and flow characteristics of the BERTIN & Cie heat exchanger. The essential features of the heat exchange module are the use of jet impingement heat transfer to maximize overall heat transfer rates which must yield minimal heat transfer material areas.

A thermal equilibrium mathematical model is set up on this basis to optimize the module efficiency and pressure drops as a function of :

- jet impingement heat transfer characteristics (slot spacing, slot distance to heat transfer surface, module length),
- hot and cold gas flow characteristics (mass flow rates, temperature and pressure levels),
- ceramic material characteristics (thermal conductivity and minimum thickness compatible with mechanical design constraints).

Optimal design leads to the following conclusions for SiC made core :

- length	: 0.340 m
- width	: 0.250 m
- height	: 0.080 m
- material thickness	: around 0.003 m
- module efficiency	: from 0.2 to 0.3
- module NTU	: from 0.2 to 0.3
- overall module transfer coefficient	: around 150 W/m^2 °C
- module pressure drop (air side)	: 200 to 300 Pascal

The third phase is about to be completed : a joint design study with Société CERAVER ends up with a self consistent ceramic design. The ceramic module is inserted within an insulating cover made of ceramic wool held by metal housing. The module has been constructed : it is now ready for testing.
The fourth Phase will consider :

i) check of mechanical integrity,

ii) steady state efficiency measurements as a function of temperature differences,

iii) conclusion on the proposed design both on technical and prospective economical grounds.

REFERENCES

(1) D. Grouset, G. Martinez, J.C. Mulet, D. Rebuffat, S. Galant, Bertin Report N° 81 Cg 67 (1981)

(2) D. Grouset, G. Martinez, J.C. Mulet, D. Rebuffat, S. Galant and al., Bertin Report N° 82 Cg 23 (1982)

(3) J.J. Cleveland, J.M. Gonzalez, K.H. Kohnken, DOE Report, DOE/05, 40174-T2 (1980)

(4) J.Y. Deysson, S. Galant, B. Mauconduit, A. Potiron, Bertin Report N° 81 Ce 02 (1981)

(5) M. Minjolle, CERAVER, Personal communication (1982)

(6) S. Galant, D. Grouset, G. Martinez, N. Martin-Neuville, D. Rebuffat, Bertin Report N° 81 Cg 75 (1981)

DISCUSSION

Question - Mr. Mischiatti, FBM/TEKSID : Concerning the ceramic recuperator development model, what is the maximum pressure difference between the two fluids ?

Answer It hat reached 0.5 bar.

Question - Dr. Ralph, UKAEA Harwell : Impingement cooling or heating is normally associated with relatively large changes of local temperature close to the impingement centre. Have you considered possible thermal fatigue problems due to such temperature gradients ?

Answer The optimization study covered both thermal and mechanical constraints. In particular, geometrical parameter values which maximize heat transfer rates are not the ones which minimize temperature gradients within the heat transfer surface. We took account of this mechanical constraint to end up with designs where temperature differences smaller than 5°C are observed in the direction perpendicular to that of the main flow. This point has been checked in the so-called 'cold' (i.e. metal) experiments.

Question - Mr. Zegers, CEC : Did you do thermal cycling experiments with the ceramic heat exchanger ?

Answer The present programme is only aimed at showing the steady state feasibility of the heat exchanger. The next development step will of course deal with this problem, which is not as severe for fixed industrial installations as it is for moving gas turbines, for instance.

Question - Dr. Pilavachi, CEC : Are the ceramic perforated plates to create gas impingement justified ? In other words, how do you compare the heat transfer area of two heat exchangers which have the same performance, one equipped with perforated plates and the other without ?

Answer Impingement heat transfer coeeficients are known to be higher than flat plate ones, for the same pressure drop and with the same temperature gradient within the boundary layer. We are therefore sure that for convective modes only, the heat exchanger length should be less. However, we are going to test the necessity of incorporating one or two perforated plates : indeed model calculations show that there exists a trade-off temperature below which only one plate is efficient enough.

Question - Dr. Pilavachi, CEC : Can you clean your heat exchanger easily ?

Answer Calculations are being carried out to check plugging properties at particulate concentrations compatible with standard French pollution regulations.

Question - Chairman : What is the likely cost of the ceramic heat exchanger ?

Answer Of the order of FF 1700/kW (260 ECU/kW). This makes it competitive with commercially available ceramic heat exchangers of more conventional structure.

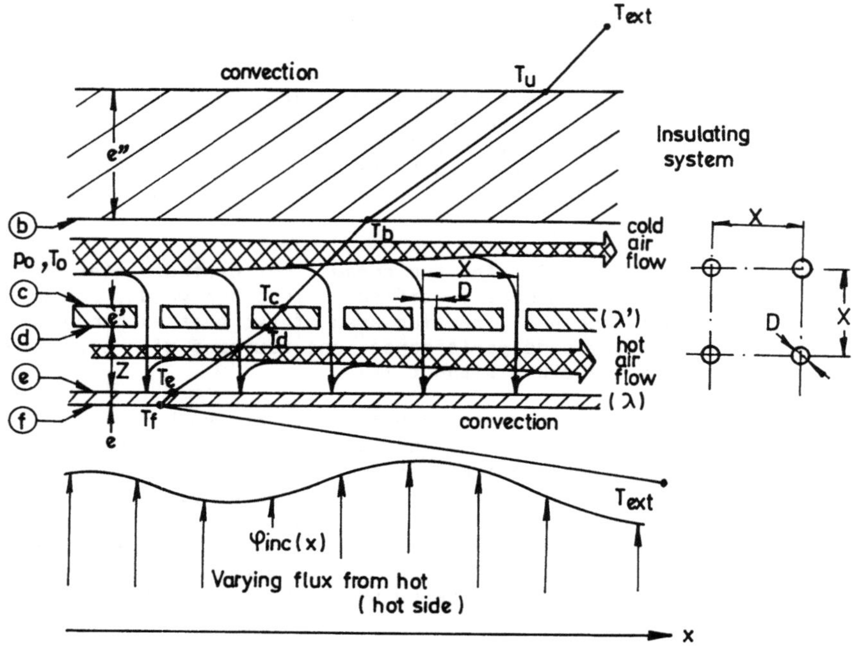

Fig 1 – The ceramic heat exchanger principles

Figure 2

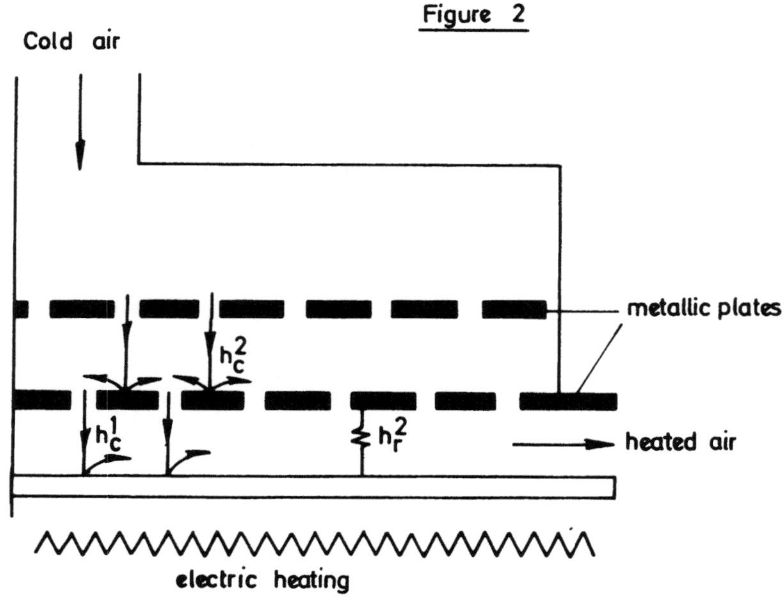

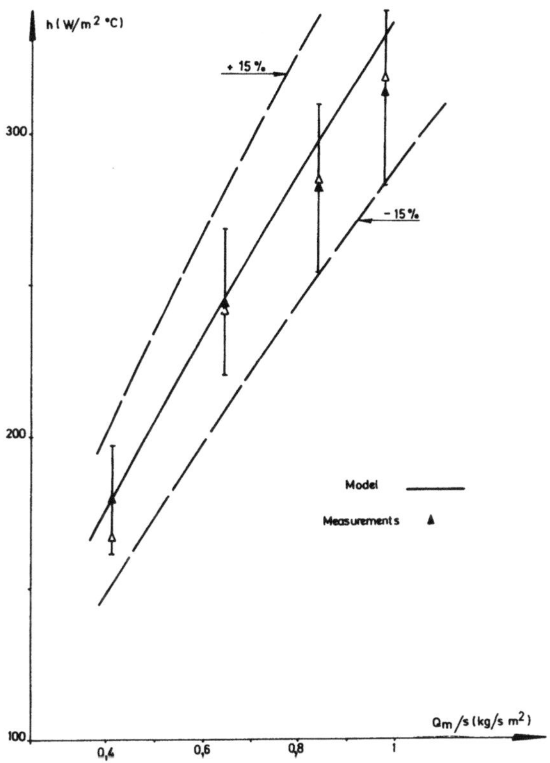
h (W/m² °C)
300
200
100
+ 15 %
- 15 %
Model
Measurements
0,4
0,6
0,8
1
Qm/s (kg/s m²)

Fig . 3

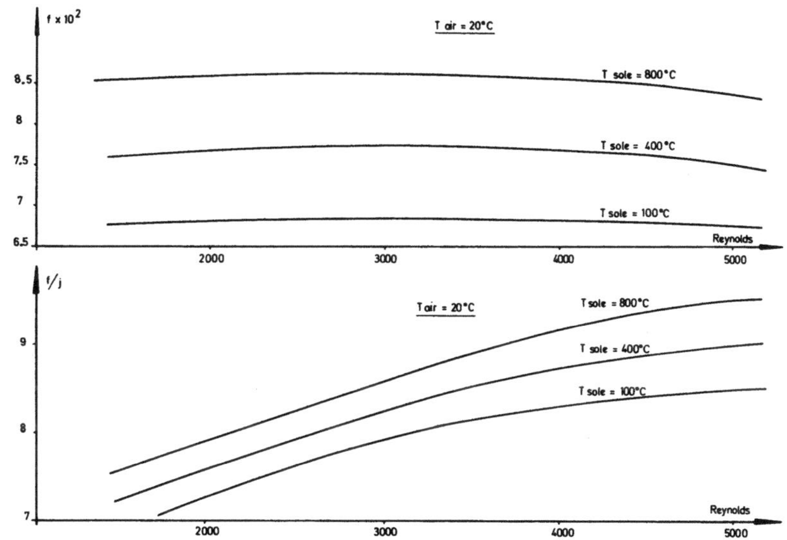
f x 10²
T air = 20°C
T sole = 800°C
T sole = 400°C
T sole = 100°C
8,5
8
7,5
7
6,5
Reynolds
2000
3000
4000
5000
f/j
T air = 20°C
T sole = 800°C
T sole = 400°C
T sole = 100°C
9
8
7
Reynolds
2000
3000
4000
5000

Fig . 4

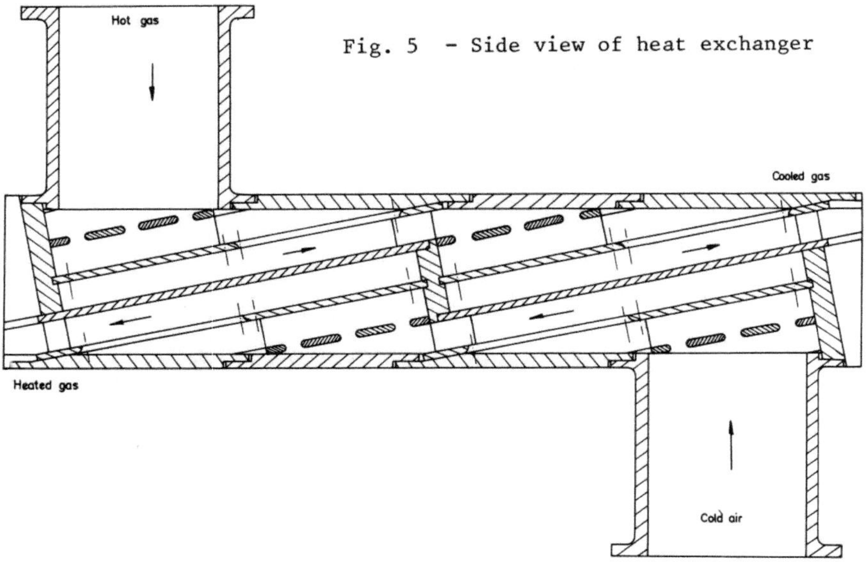

Fig. 5 - Side view of heat exchanger

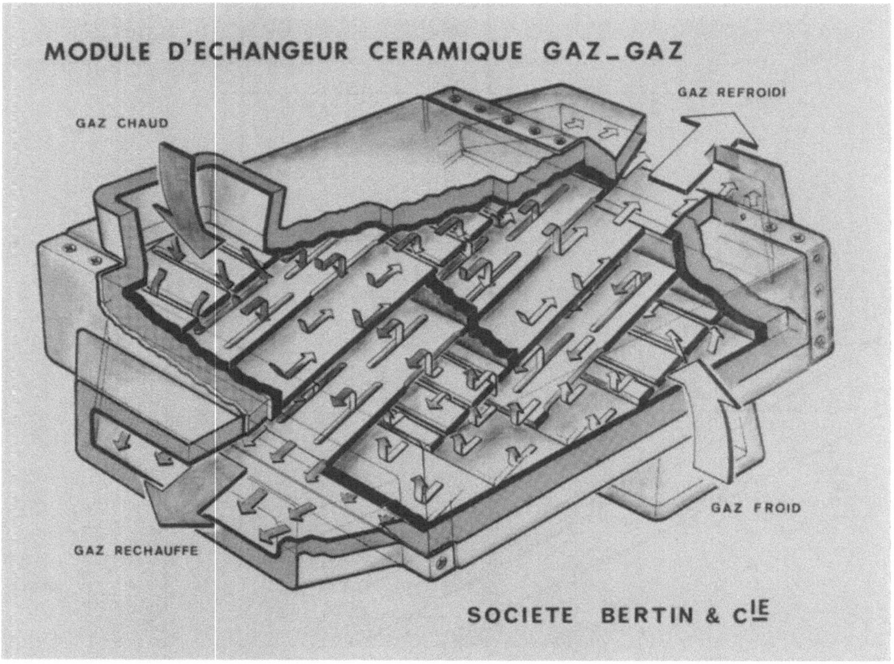

Fig. 6

Fig 7_ Schematic of experimental test rig

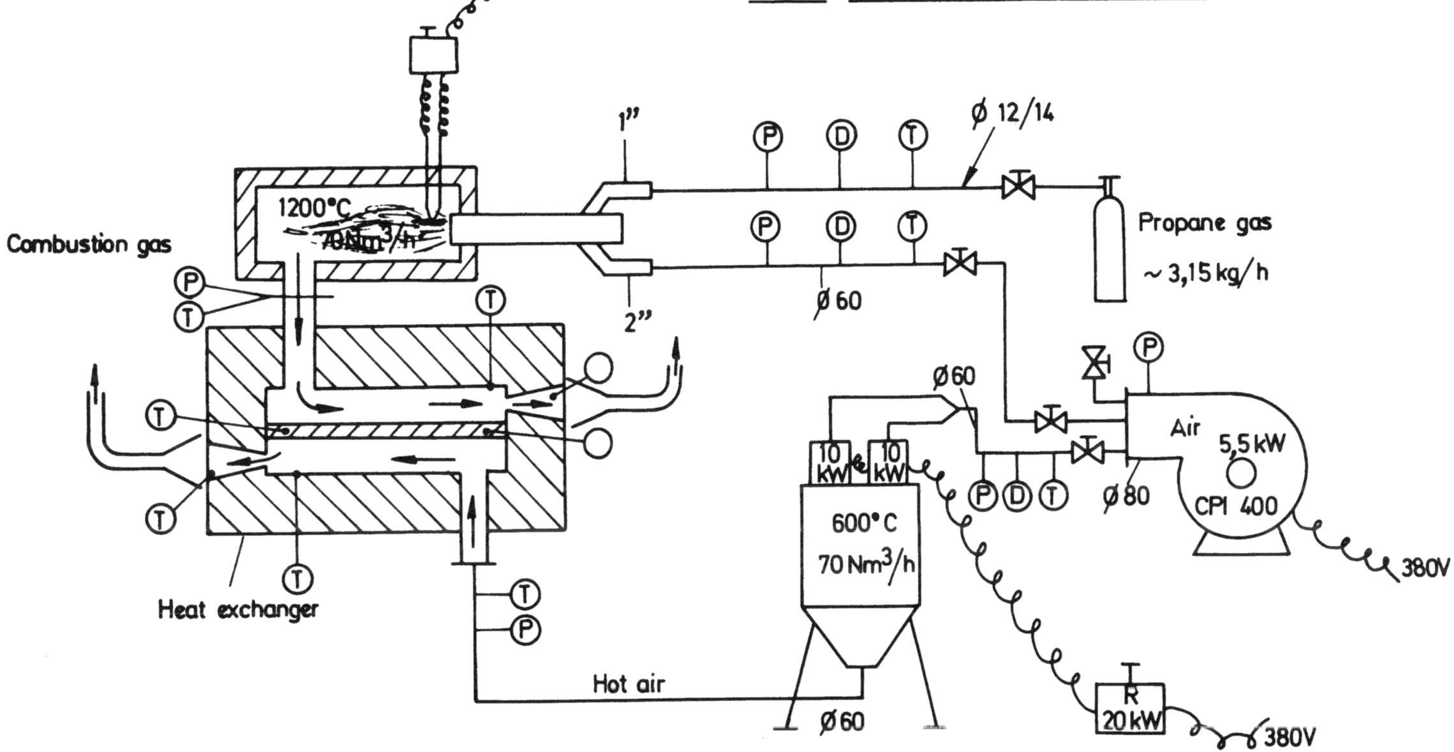

COMPACT GAS-GAS HEAT EXCHANGER

Authors : A. GREHIER, C. RAIMBAULT, A. ROJEY (IFP)
B. CHLIQUE, J. DREUILHE (CREUSOT-LOIRE)

Contract number : EE-B-1-116-F(G)

Duration : 18 months 1 June 1981 - 31 December 1982

Head of project : A. Rojey, Institut Français du Pétrole

Contractor : Institut Français du Pétrole

Adress : 1 et 4 Avenue de Bois Préau
92502 RUEIL-MALMAISON (France)

Summary

The european market study carried out by IFP and Creusot-Loire in the present framework has shown that a 2 millions tOE/year savings can be expected from new gas-gas heat recovery systems in the present industrial context. The main constraint to be overcome is the cost of setting up in industrial plants. The compactness and modular conception of the new stacked perforated plates (SPP) gas-gas heat exchanger will undoubtedly help in getting this new equipment on to the market The exchanger consists of a block formed by the stacking of judiciously superpositioned perforated metal plates whose apertures create the fluid flow passages. This compact gas-gas cross flow heat exchanger is easy to manufacture, with low pressure drops and easy to clean fumes passages. The different plate perforation geometries, tested in laboratories on small models (exchange area $\sim$ 0.35 m^2) showed the possibility of attaining specific areas of 125 m^2/m^3 and heat transfer coefficients up to 70 W/m^2 °K. The data obtained from these tests serve as a basis for planning industrial devices and have been used to design a prototype (exchange area : 12.5 m^2) which includes cleaning devices. The tests of this prototype will be carried out on a 1.6 MW boiler with fumes flowrates in the range of 2000 m^3/h and outlet temperatures of up to 400 °C.

1.1 Introduction

Rising oil costs and energy conservation considerations have led to an increasing demand for the design of more efficient heat exchange equipment. Heat recovery from fumes is particularly attractive because it conserves energy each time combustion occurs in various industry and heating operations. According to a European market study carried out by IFP and Creusot-Loire in the present EEC framework, the minimum potential demand of industrial gas-gas heat recovery systems to be installed in the present industrial context, would result in a savings of about 2 millions tOE a year broken up as shown Table I.

Industrial area	tOE
Refining	100 000
Chemistry	250 000
Steel industry	390 000
Paper industry	800 000
Food industry	450 000
Total	1 990 000

TABLE I

To these savings should be added recoveries from the glass, metallurgy and textile industries for which detailed data are not available.

On the other hand, this market survey shows that the constraints which have to be overcome to enable such new equipment to be marketed are mainly related to the space they require and to their installation cost. The choice of the heat recovery system will often depend on its positionning possibilities and on the time required to set it up. As such, it is worth remembering that the cost of an installed gas-gas heat recovery system can be anything between 1.5 and 4 times the cost of the basic heat exchanger itself.

It is expected that the SPP gas-gas heat exchanger will cope with these environmental and economic constraints due to its compactness and modular conception resulting in a high installation flexibility. The descriptions given hereafter are followed by a presentation of the tests and results we obtained on small models in laboratories during the first phase of our work schedule which also includes, the design and the construction of a prototype (2nd phase) and the test of the latter on an industrial boiler (3rd phase).

1.2 Principle (1), (2)

The SPP heat exchanger consists in a block formed by stacking of perforated metal plates so as to obtain channels by superposition of the perforations. Two stacking modes have been tested. In the first one(hereafter defined as "staggered/staggered mode") each perforation of each plate communicates with two perforations of the preceding plate and with two perforations of the following plate (Figure 1).

In the second mode ("staggered/lined up" mode) rows of staggered perforations, alternate with rows whose perforations are simply lined up.

This second arrangement is obtained by alternately stacking plates having aligned perforations and plates having staggered perforations (Figure 2). Besides, a cross-flow heat exchanger is easily obtained by a judicious cutting up of the plates and by sealing 1 row out of 2 on the end plates. Figures 1 and 2 show cross-sections of air and fumes channels in both cases described above. The second one offers, in addition, the benefit of straight channels which are especially appreciated in cases of heat recovery from polluted gas. A perspective view of such an exchanger is shown Figure 3.

1.3 Experimental devices

A schematic diagram of the device used for these tests is given in Figure 4. Before being introduced into the heat exchanger, (consisting of a cubic stack, 15x15x15 cm), the flow rates of cold and hot fluids are measured in ambient conditions. In addition, the experimental device includes mean pressure and temperature mesurements for the two fluids as well as the heating system for one of them. A thermal insulation reduces external heat loss around the exchanger. Hot air is fed perpendicularly to the plane of the plates and cold air is circulated in a direction parallel to these planes and to the direction of the alignment of the perforations (Figures 5 and 6). A general view of the experimental device is shown in Figure 7.

1.4 Physical features of the exchangers tested

Two types of heat exchangers have been investigated using plates with oblong perforations stacked according to the two modes mentioned above. We have also investigated a third exchanger using plates perforated with square perforations stacked according to the "staggered/lined up" mode. Physical featurs of these exchangers are given on Figure 8.

1.5 Tests

The main purpose of the tests was to establish the relationship linking heat transfer coefficients and pressure drops with cold and hot gas velocities for three types of stacks investigated.

The results obtained with the "SPP2" and "SPP3" heat exchangers stacked according to the "staggered/lined" mode are given here-under.

1.5.1 Heat transfer coefficient

As an example, experimental data concerning the SPP2 staggered/lined up stacking mode are given in Table II. Measurements of heat released from the hot air are greater than those received by the cold air. As no significant leaks (< 1 %) were detected between the two compartments of the exchanger, this thermal balance defect could only be explained by thermal losses from the exchanger itself to the ambient air. A corrected heat transfer coefficient value has been calculated by assuming that these losses are respectively proportional to differences between the temperatures of the cold and hot air and the ambient air temperature.
From this experimental data we can establish the relationship between heat transfer coefficients and cold air mass velocities, plotted in Figure 9, which can be expressed as :

$h_1 = 15.2\ (G_1)^{0.67}$ h : heat transfer coefficient

and $h_2 = 12.6\ (G_2)^{0.55}$ G : mass velocity

respectively for the "SPP2" and "SPP3" stacked plates. It is interesting

TABLE II

STAGGERED/LINED UP SPP 2 HEAT EXCHANGER*

Cold air mass flow rate (kg/s).10^{-3}	32.7	32.7	29.0	29.0	23.1	23.1	19.3	19.3	10.0	10.0	3.24	3.24
Hot air mass flow rate (kg/s).10^{-3}	32.7	32.7	29.0	29.0	23.1	23.1	19.3	19.3	10.0	10.0	3.24	3.24
Cold air mass velocity (kg/m^2.S)	9.28	9.28	8.21	8.21	6.56	6.56	5.46	5.46	2.20	2.20	0.71	0.71
Hot air mass velocity (kg/m^2.S)	10.17	10.17	9.00	9.00	7.19	7.19	5.99	5.99	2.41	2.41	0.78	0.78
Inlet cold air temperature (°C)	69.3	72.2	74.0	76.0	81.9	84.8	89.5	90.0	25.5	24.9	28.6	27.2
Outlet cold air temperature (°C)	80.1	99.1	83.5	101.5	88.5	107.3	93.9	109.9	57.3	78.1	60.0	84.3
Inlet hot air temperature (°C)	97.2	142.4	99.2	141.9	101.2	145.5	104.2	142.8	100.4	152.3	97.4	147.9
Outlet hot air temperature (°C)	84.9	113.1	88.1	113.8	92.0	119.0	96.1	119.3	64.2	88.2	56.4	83.5
Pressure drop on cold air flow (Pa)	2310	2310	1630	1900	1090	1360	950	1220	210	200	40	50
Pressure drop on hot air flow (Pa)	230	220	180	200	110	120	80	100	< 20	20	< 10	< 10
Cold air velocity (m/s)	9.14	9.43	8.18	8.41	6.66	6.87	5.64	5.77	1.96	2.02	0.64	0.66
Hot air velocity (m/s)	10.49	11.55	9.35	10.22	7.53	8.26	6.32	6.85	2.42	2.21	0.77	0.86
Heat transferred (from cold air Δt) (W)	353	879	276	739	152	519	85	384	318	532	102	187
Heat transferred (from hot air Δt) (W)	402	958	322	815	212	612	156	454	362	641	132	211
Corrected heat transferred (W)	374	909	297	768	179	557	119	413	329	558	111	193
Heat transfer coefficients (W/m^2.°K)												
- From cold air Δt	68.1	67.0	59.4	60.6	43.0	46.0	32.7	39.6	24.9	24.8	10.1	10.0
- From hot air Δt	77.5	73.0	69.4	66.8	59.9	54.3	60.3	46.8	28.4	30.0	13.1	11.3
- Corrected	72.1	69.3	63.9	62.9	50.8	49.3	46.2	42.6	25.8	26.1	11.0	10.3

* Mean cold air duct cross section : 35.25 . 10^{-4} m^2
Mean hot air duct cross section : 32.15 . 10^{-4} m^2
Exchange area = 0.339 m^2

to point out that these equations are very similar to the ones deduced from the formula set up by DONOHUE and KERN (3), (4), (5) for shell and tube heat exchangers.

1.5.2 Pressure drops

The linear relation deduced from the experimental data between the pressure drops and the kinetic energies of each fluid shows the prevailing influence of dynamic losses. The experimental relation relative to the air flowrates can be expressed as :

$$\Delta p = K \frac{\rho V^2}{2}$$

Δp : pressure drop

V : air velocity in the mean air cross section

K values obtained for both exchangers are given in Table III.

	K Values $\Delta p = K \frac{\rho V^2}{2}$	
	"SPP3"	"SPP2"
Cold air side	41.4	55.0
Hot air side	3.2	4.3
Length of cold and hot air ducts : 0.15 m		

TABLE III

If a pressure drop of 1 1/2 times the kinetic energy at the inlet and outlet of the air ducts is assumed, the general pressure drop velocity relation shown in Table III, becomes, for any air duct length L :

$$\Delta p = \left[K'L + 1.5\right] \frac{\rho V^2}{2}$$

K' values calculated for the two types of perforated plates are given in Table IV.

	K' Values $\Delta p = \left[K'L + 1.5\right] \frac{\rho V^2}{2}$	
	"SPP3"	"SPP2"
Col air side	266	357
Hot air side	11.1	18.4

TABLE IV

These data set ut the basis for the following definition of the prototype exchanger.

2.1 Industrial design

After the first serie of tests carried out by IFP, we already examined the problems involved in industrial application of an SPP heat exchanger.

They concern mainly the devices used to :

- Obtain a good compactness, i.e., avoid the swelling of plates during stacking, so as to minimise the leakage between gas flows which would be prejudicial to heat exchanger performance.

- Allow the cleaning of hot gas channels, this gas being able in some cases to carry dust.

- Obtain a good resistance to corrosion, according to gas properties.

To solve these problems, special attention must be given to :

- The choice of the plate material.

- The choice of drilling and the arrangement of stacked plates.

- The surface condition and tolerances which must be obtained for the plates. Precise specifications must be established in agreement with the supplier.

2.2 Choice of plates

We have at first selected two types of material for SPP :

- Perforated plates made of Alu. alloy (type 5754 for instance). A lighter and compact apparatus allows a greater flexibility of layout.

- Perforated plates made of stainless steel (type 316 for instance). This material offers a wider field of application, concerning temperature and corrosion.

For a first construction, these two types of materials would make easier the stacking of perforated plates :

- Alu. alloy, thanks to its greater ductility.

- Stainless steel, thanks to its cold-creep characteristics which allow to obtain and maintain plane plates by means of pressure

Other materials such as carbon steel would be considered later, the aim being lower cost. These materials will benefit from the experience acquired with the previous ones.

The choice of the type of drilling results from a compromise between :

- The maximum compactness of the SPP which leads to numerous perforations of small size.

- A reasonable pressure loss.

- The technological feasibility of the SPP (tightening of the plates, effect of tolerances...).

Among the existing perforated plates, we have chosen plates with oblong perforations. This type of perforation, together with a staggered/lined up arrangement, gives hot channels of constant section and we can reasonably expect to clean them by a blowing device.

In order to make easier the stacking of the plates, we prescribe to the supplier to furnish the plates with a smooth surface : we ask for a chemical fettling. It is likely that, later, this operation will be reduced to an ordinary brushing of the plates, especially where Alu. Alloy plates are concerned. Further, for each plane, the orientation of the plane during drilling will be marked.

2.3 SPP design

With the plates defined above, we designed a SPP heat-exchanger unit corresponding to the following thermal characteristics :

Hot gas flow	12.300 kg/h
air flow	11.680 Kg/h
Hot gas temperature	400 to 230 °C
air temperature	20 to 213 °C
Thermal output	650 kW
Pressure drop (air and gas)	10 mbar

By extrapolating results of IFP tests, we obtained the following dimensions :

Length		2 meter
Width		0.480 m
Height		0.740 m
Weight	∿	3.5 t (stainless steel)
	∿	1.6 t (Alu.alloy)

2.3.1 Main features

- Tightening of the plates

Stacking being achieved, the tightening of the plates is realized by means of flat bars distributed in some hot channels and higher than the stacking. Those flat bars contain at their ends pieces of square section pinned to the bars. These sections are threaded, thus allowing to tighten the plates by screws bearing on transverse flat bars, these bars distribute the load along the plates. This device makes it possible to adjust the tightening load during the assembling of the SPP.

- Cleaning device

The planned cleaning device is a steam blowing type. The steam is distributed in a pipe having the same length as the plane, this pipe is fitted with nozzles, each nozzle corresponding to a row of hot channels. At each end of the pipe is welded a threaded rod, connected to a mechanical device. Thus, it is possible to translate the pipe during the cleaning operations, and the nozzles discharge in the various hot channels of each row, one after another. When not operating, the pipe is removed of the hot gas flow.

2.3.2 Solaize prototype

The SPP prototype which will be erected at Solaize (IFP development center), is an element of the above unit, and will be made of Alu. alloy. It is adapted to the boiler characteristics:

Gas flow	2000 Nm^3/h
Boiler exit temperature	280 °C

This temperature could be raised up to 400 °C by by-passing parts of the boiler bundle.

This prototype consists of 185 stacked plates, the dimensions of each plate being 370 x 480 mm.

Tests and measures that will be carried out with this prototype have the following aims :

- To check there is no appreciable leakage between gas and air flows.

- To verify the SPP heat exchanger performance with representative operating conditions.

- To test the efficiency of the cleaning device.

- To obtain data concerning the behavior of light alloy at these operating conditions.

3.1 Conclusion

This study showed that gas gas heat recovery systems offer a potential savings of 2 millions tOE/year to which the new SPP gas gas heat exchanger could very well contribute. Indeed, its modular conception could provide answer to the geometric constraints which are often required of new devices to be installed in a given industrial system.

In addition laboratory tests determined the specific exchange and pressure drop features of several types of SPP exchangers according to different perforated plates and stacking modes. Heat exchange coefficients up to 70 $W/m^2.°K$ were obtained in the range of air velocities of 10 m/s with corresponding pressure drops lower than 2500 Pa. These tests also determined the relation linking the heat exchange coefficient and the pressure drop to the air velocity. These data led to the establishment a of scaling up program which has been used to determine a development prototype (exchange surface area : 12 m^2) and constituted the basis of a modeling study now in progress.

The research carried out shows the feasibility of the SPP type heat exchanger, at satisfactory economical conditions. If the tests carried out in Solaize confirm the expected performances, it will be possible to go on further and design an industrial size unit.

References

(1) ROJEY A., COHEN G.
Brevet d'invention N° d'enregistrement national 79/11529

(2) ROJEY A.
Brevet d'invention N° d'enregistrement national 81/03902

(3) WUITHIER P.
Le Pétrole, raffinage et génie chimique Tome II, pp 1033-1141, Ed. Technip, Paris 1965

(4) DRUMOND W.
The design of shell and tube heat exchanger. Heat exchangers, Von Karman Inst. Lecture Series, AGARD, LS-57-72, 1972

(5) AFGAN V. and SCHLUNDER E.U.
Heat exchangers, design and theory, source book, Mc Graw-Hill, 1974

DISCUSSION

Question - Dr. Pilavachi, CEC : Is the pressure drop through this heat exchanger acceptable ?

Answer The maximum pressure drop is taken as a constraint in the design of the heat exchanger. This maximum pressure drop must be small for an exchanger recovering heat from stack gases, (around 1000 Pa for a large industrial air preheater). Admitting larger pressure drops would lead to larger specific heat exchanger areas.

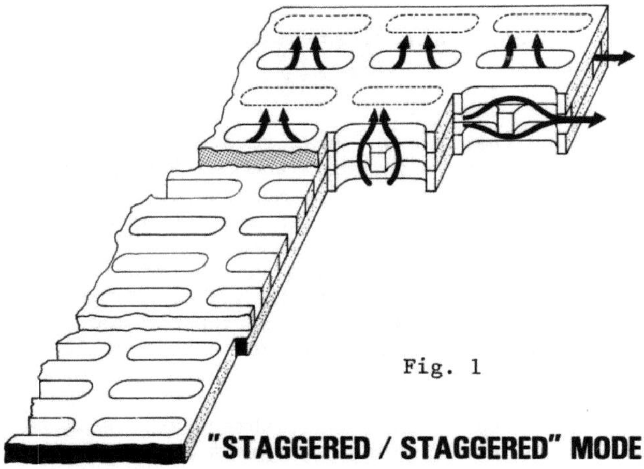

Fig. 1

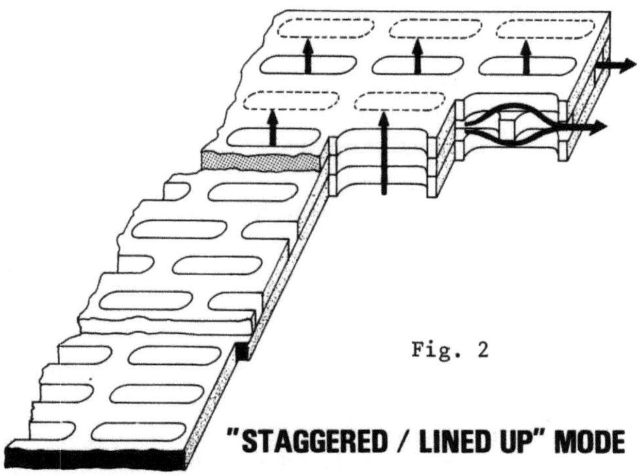

Fig. 2

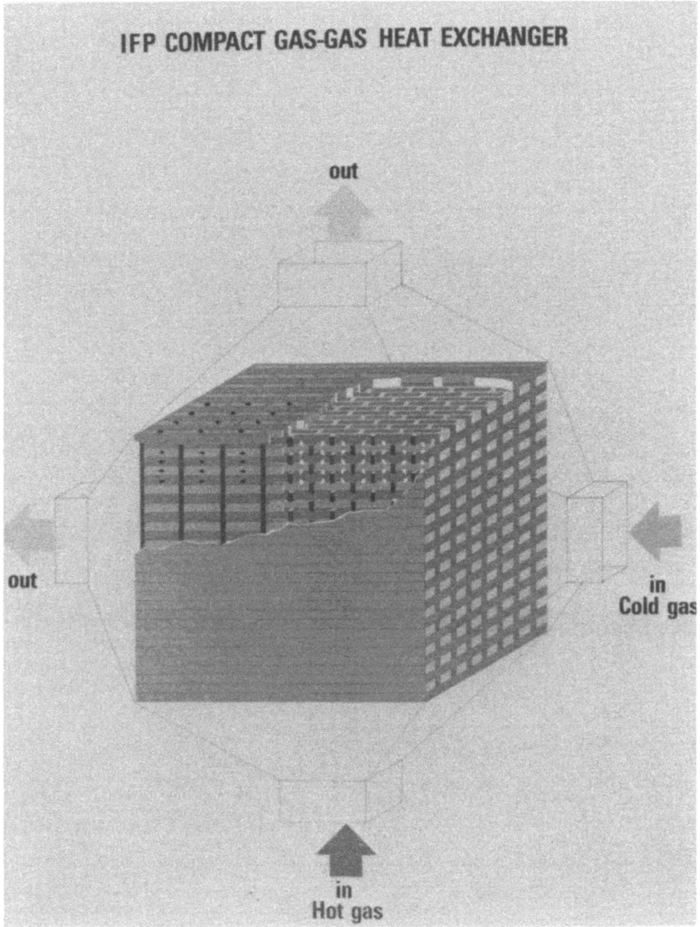
IFP COMPACT GAS-GAS HEAT EXCHANGER
out
out
in
Cold gas
in
Hot gas

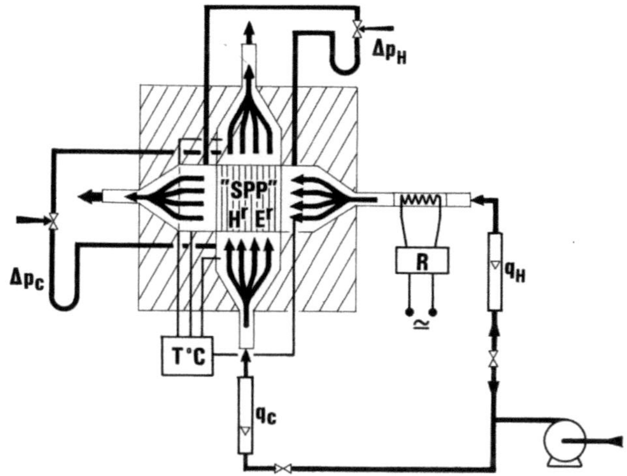
Δp_H
Δp_C
"SPP"
H^r E^r
R
T°C
q_H
q_C

↑ Fig. 4

← Fig. 3

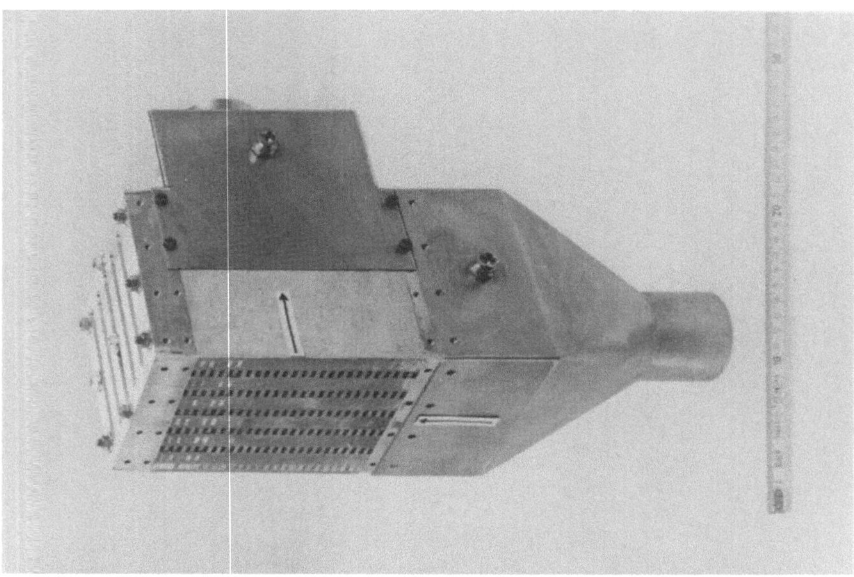

Fig. 5

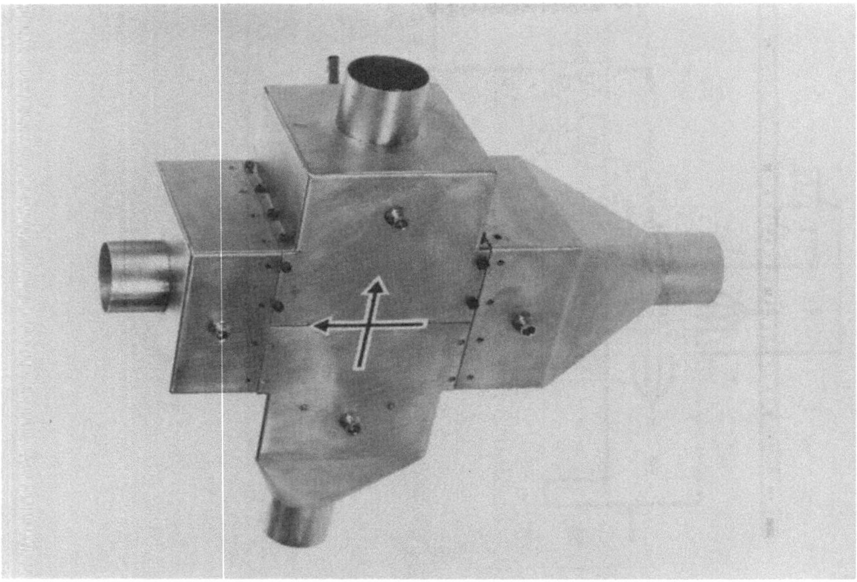

Fig. 6

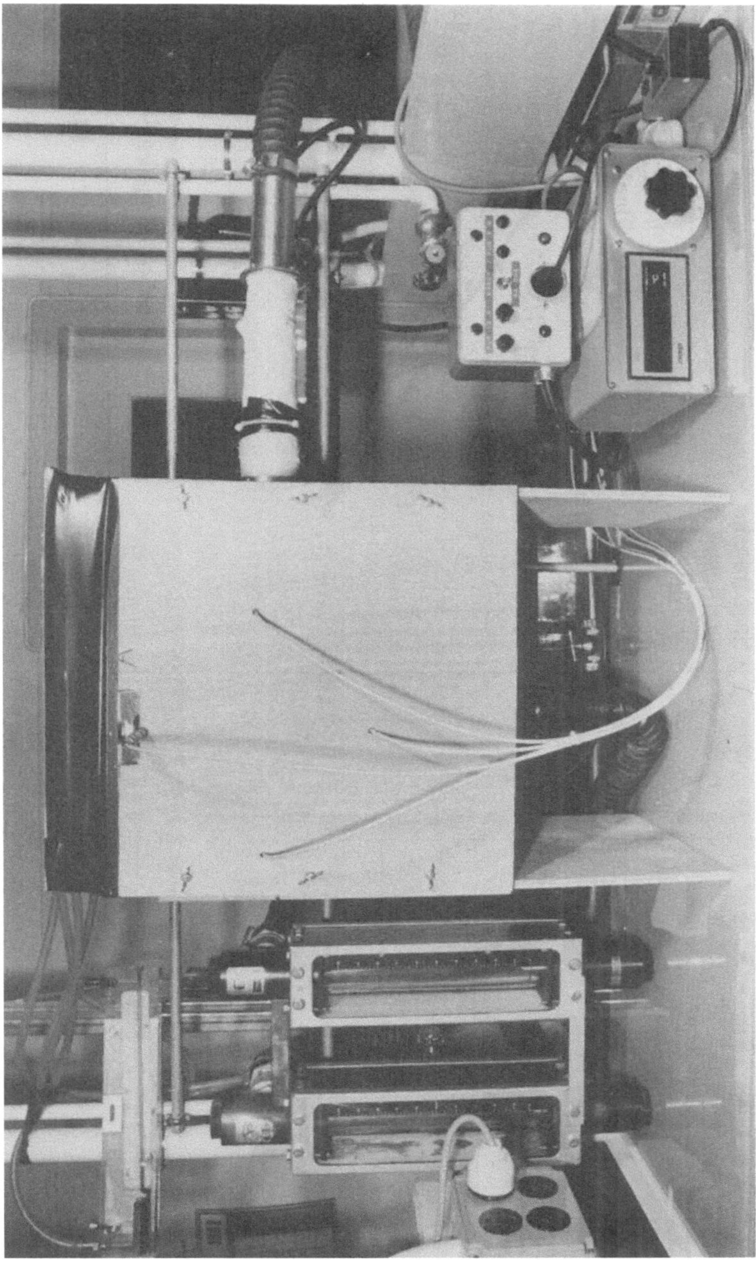

Fig. 7

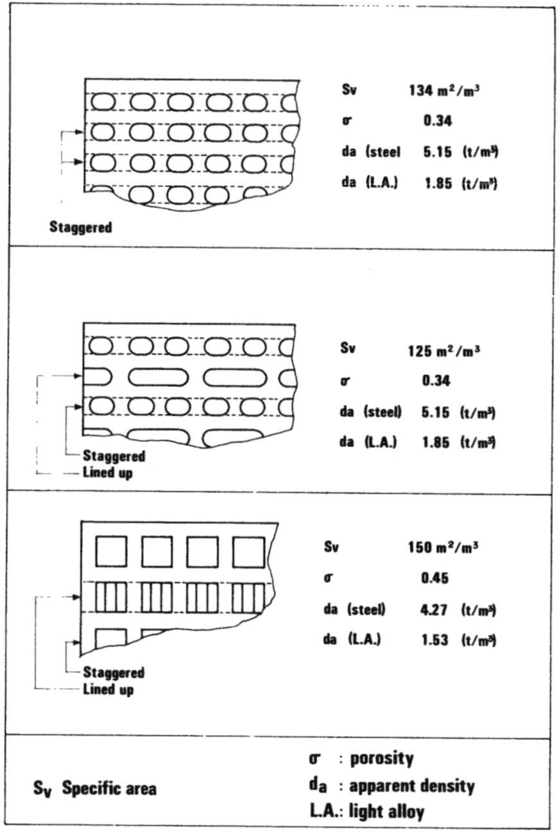

Sv 134 m²/m³
σ 0.34
da (steel 5.15 (t/m³)
da (L.A.) 1.85 (t/m³)
Staggered
Sv 125 m²/m³
σ 0.34
da (steel) 5.15 (t/m³)
da (L.A.) 1.85 (t/m³)
Staggered
Lined up
Sv 150 m²/m³
σ 0.45
da (steel) 4.27 (t/m³)
da (L.A.) 1.53 (t/m³)
Staggered
Lined up
Sv Specific area
σ : porosity
da : apparent density
L.A.: light alloy

Fig. 8

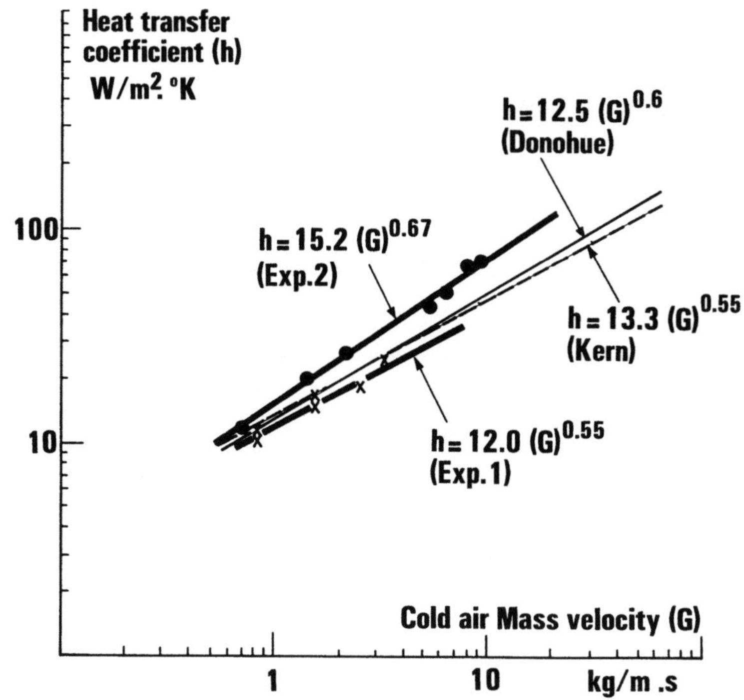

Heat transfer coefficient (h) W/m². °K
100
10
h=12.5 (G)^0.6 (Donohue)
h=15.2 (G)^0.67 (Exp.2)
h=13.3 (G)^0.55 (Kern)
h=12.0 (G)^0.55 (Exp.1)
Cold air Mass velocity (G)
1
10
kg/m .s

Fig. 9

THE EVAPORATION OF VISCOSE PROCESS LIQUORS BY LOW ENERGY MEANS

Author : R. Thornton

Contract Number : EEB-1-101-80-UK (H)

Duration : 34 Months 1st June 1980 - 31st March 1983

Head of Project : Mr. B.S. March (Head of CDEC)
Dr. R. Thornton (Project Manager)

Contractor : Courtaulds PLC

Address : Courtaulds PLC
Chemical and Development Engineering Centre
P.O. Box 16
Foleshill Road
Coventry CV6 5AE
United Kingdom

Summary

A research programme leading to the development of a low energy process for the evaporation of water from viscose process liquors is in progress. The process employs mechanical vapour recompression techniques combined with the use of thin walled plastic tubes as a heat transfer surface. Laboratory investigations have been undertaken to determine the physical properties of the liquors involved, and data on viscosities, specific gravities, boiling point elevations, thermal conductivities and specific heat have been gathered. An examination of the properties of plastic tubes has been executed, with particular emphasis on short and long term mechanical stability, resistance to fouling and wettability. A bench scale mechanical vapour recompression evaporator has been constructed; heat transfer coefficients of 800 - 1000 $W\ M^{-2}\ {}^{o}K^{-1}$ across the tube bundle have been obtained with the viscose liquors.
A research plant to evaporate 100 - 150 kg hr^{-1} water has been designed and construction is in progress. The detailed mechanical design has produced a number of interesting engineering challenges, particularly with regard to the compressor, the concrete vacuum vessel and its corrosion resistant lining and the tube bundle. A strategy to meet these challenges has been formulated.

1. Introduction

Courtaulds PLC believes that the long term outlook for viscose fibre production must be favourable: the raw material, wood pulp, is a renewable biological resource. It is clearly important to minimise the energy requirement for fibre production. One of the areas of highest energy consumption in viscose fibre plants is the evaporation of water from process liquors prior to their recycling. It is towards reducing this energy consumption that the current work is directed. In 1980 it was calculated that if this work were successful and the resulting technology applied to 50% of the viscose production in the EEC (apart from Courtaulds) the energy saving would be around 1.7×10^6 GJ/year.

The major innovation upon which the project rests is the use of thin walled plastic tubes as a heat transfer surface in a mechanical vapour recompression evaporator. Because the heat transfer surface can be cheap, very large areas may be employed with correspondingly low temperature driving forces across them. This leads both to a process which is thermodynamically very efficient, and also to a simplified and hence less expensive form of compressor.

The essential features of the mechanical vapour recompression evaporator are depicted in Figure 1. Viscose process liquors are distributed over the outside of a bundle of thin walled plastic tubes (1). Boiling occurs on the tubes, and the concentrated liquor is collected underneath the bundle (2). The water vapour from the boiling liquor passes through a bank of demisters (3) and into a compressor (4). Here the pressure of the vapour is raised so that its condensing temperature is above the temperature of the boiling process liquors. The vapour condenses inside the plastic tubes, supplying latent heat to the liquors outside. The condensate is collected (5) and pumped away. A purge pump (6) removes any incondensible gases from the tubes and maintains a suitable pressure inside the containment vessel (7). This pressure is such that the process liquors boil at the temperature of the incoming stream. Thus all liquid streams enter and leave the evaporator at the same temperature, and so no external heat exchangers are required.

The objective of the current research programme is to produce a design for a commercial evaporator, working on these principles, with a very low specific energy consumption. It is envisaged that this research programme will last some 34 months, and it is divided into three phases. The first of these was to carry out laboratory investigations into the physical properties of the process and constructional materials, and also to build a bench scale evaporator in glassware. This work is now completed. The second phase is largely concerned with the construction of a small research plant at a viscose production site, and the third phase is devoted to a study of the results of the programme and a detailed technical, economic and commercial assessment of the process.

At the present time we are a part of the way through the second phase. The construction of the research plant is underway, and completion is scheduled for this autumn.

2. Laboratory Investigations

The laboratory investigations fall into two areas. The first of these has been concerned with the properties of the viscose process liquors, and the second with the behaviour of the plastic tubes.

2.1 Viscose Process Liquors

The relevant physical properties of a number of process liquors have been determined in the temperature range 45 - 55°C. A summary of the results is given in Table 1.

Property	Temperature °C	Value	Units
Viscosity	45 - 55	1.38×10^{-2} to 2.93×10^{-2}	gm cm^{-1} sec^{-1}
Density	45 - 55	1.273 to 1.354	gm cm^{-3}
Elevation of Boiling Point	corresponding to 80mm mercury	2.90 to 3.95	°C
Thermal Conductivity	50°C	5.2×10^{-3} to 7.2×10^{-3}	W cm^{-1} $°C^{-1}$
Specific Heat	50°C	3.20 to 3.31	J gm^{-1} $°C^{-1}$

Table 1

Properties of Viscose Process Liquors

2.2 Plastic Tubes

The material that has been identified as being suitable for the heat transfer surface is a cross-linkable polyethylene. The cross-linking is effected by exposing the polymer, which contains reactive groups, to water. The tube for this project is produced in layflat from. Both high and low density versions of the polymer can be made, and although the low density material is good enough for the current project, the high density polymer has many superior properties. Unfortunately it has proved very much more difficult to produce uniform layflat tube from the high density material.

2.2.1 Rates of Cross-Linking

Normally a catalyst is incorporated into the material to speed up the rate of cross-linking on exposure to water. However, it was determined that the presence of the catalyst adversely affected the wettability of the plastic surface. Moreover, the viscose process liquors act as their own catalyst. As the cross-linking process will be carried out after assembly of a tube bundle, it is important to know the rates of cross-linking under various conditions. It was found that 60% cross-linking of the polymer would be obtained in 60 - 200 hours according to the temperatures employed.

2.2.2 Wettability of Plastic Tubes

In order to achieve good liquor distribution over the tube bundle in an evaporator, it is necessary to ensure that the plastic surface is wetted by the liquors. Laboratory experiments showed that the process of cross-linking the polymer by viscose process liquors in the absence of catalyst produced a surface which was markedly more wettable than conventional plastics.

Moreover, long term tests indicate that the effect is permanent. The wettability (or wetting tension) of the layflat tube after treatment in process liquor was found to be as high as 72 dynes cm^{-1}.

2.2.3 Deposition of Salts

Experiments were initiated to investigate whether crystals of inorganic salts deposited from process liquors would stick to the plastic tubes. The work was carried out under conditions of heat transfer through the wall of the tube. A control standard conventional evaporator tube was also employed. It was found that although the standard tube became heavily fouled with crystals, no crystals adhered to the plastic tubes. Figure 2 shows the details of the apparatus used.

2.2.4 Thermal Conductivity and Thermal Expansion

The thermal conductivities of high and low density polymer were measured at 0.61 $W\ m^{-1}\ {}^{o}K^{-1}$ and 0.33 $W\ m^{-1}\ {}^{o}K^{-1}$ respectively. The coefficients of thermal expansion $1.6 \times 10^{-4}\ {}^{o}K^{-1}$ and $6.0 \times 10^{-4}\ {}^{o}K^{-1}$.

2.2.5 Creep and Creep Rupture

If plastic heat transfer surfaces are to be used in an industrial application it is essential that the lifetime of the tubes is adequate. In the current project the aim is for a lifetime of at least five years. This means that time dependent properties of the plastic tubes, such as creep and creep rupture, must be investigated. Photo-degradation is not important here as the tubes will be in total darkness. The Rubber and Plastic Research Association of Great Britian (RAPRA) was commissioned to carry out accelerated tests to determine the effects of creep in an environment of viscose process liquor. These tests could be carried out conveniently only in "synthetic" liquors made up in the laboratory. In order to ascertain whether trace components present in a production environment would affect the properties of the polymer, a number of spot checks were carried out in Courtaulds' own laboratories. It was found that, at the temperature and stress levels present in a vapour re-compression evaporator, the lifetime of the tubes should exceed five years. The presence of trace chemicals in commercial process liquor had no detectable effects on the performance of the polymer.

3. Bench Scale Evaporator

A bench scale vapour recompression evaporator has been constructed. This has been used to verify the principles of vapour recompression as applied to viscose process liquors, to determine overall heat transfer coefficients for the horizontal sprayed liquor evaporator tube bundle, and to look at the effects of incondensible gases in the system.

A photograph of the apparatus appears in Figure 3 and a line diagram is shown in Figure 4. Process liquor from the flask (4) is pumped by a rotary lobe pump (5) via a rotameter (6) and a heating coil (7) into a header which is connected to a number of spray bars (9). The spray bars distribute the liquor evenly over a horizontal bundle of polyethylene tubes (10) mounted within a QVF glass shell (1). The tubes are in tension between two tube plates, and vertical baffles hold the tubes and the spray bar in position along their length. The liquor boils on the outside of the tubes, the resulting vapour passing to a rotary vane compressor (2). The compressed vapour passes to the tube side of the evaporator where it condenses.

The tube side is connected to the shell side through a water lute (8) which protects the tubes against over-pressurisation. In this apparatus the concentrated liquor is returned to the flask (4) and recirculated after mixing with the condensate from the tube side. The rate at which condensate is produced can be measured by means of a graduated tube (3). This vacuum is produced and maintained by means of an oil sealed rotary vacuum pulp (15). Water vapour is removed from the air purge to the vacuum pumps by means of water-cooled condenser (11), connected to a measuring cylinder (12), and a dry ice trap (13). An oil bubbler (14) gives a visual check on the rate of air removal from the system.

This research showed that a heat transfer coefficient of 1000 $W\ m^{-2}\ {}^{o}K^{-1}$ could be achieved with an irrigation rate of 108 litres per hour per metre of tube. For the purposes of further plant design a "safe" value of the heat transfer coefficient of 800 $W\ m^{-2}\ {}^{o}K^{-1}$ will be assumed. The effect of air leakage was studied, and the results are now being correlated in terms of a mathematical model of the system. Machine computations have shown that the presence of incondensible gases in viscose process liquors (for example, hydrogen sulphide and carbon disulphide) as well as air leaks into the equipment, has an important bearing on the detailed design of any practical plant.

4. Research Plant

The next stage in this research programme is the building of a small research plant at a manufacturing site. Detailed flowsheets and process and instrumentation diagrams for this plant have been drawn, and construction has commenced.

From the outset it was determined that this research plant should fulfil a number of criteria. It should:

(i) Ensure that the process is continuously and reliably workable under normal production plant conditions.

(ii) Allow the study of long term fouling effects.

(iii) Determine the stability and durability of plastic tubes, their supports and fastenings.

(iv) Investigate the suitability of the containment vessel and its acid resistant lining.

(v) Demonstrate the operability and reliability of the compressor.

(vi) Verify assumptions regarding liquor distribution, heat transfer coefficients and other parameters.

Originally it was hoped to design a research plant with an evaporative capacity of 300 $kg\ hr^{-1}$ of water. Owing to economic constraints it has been necessary to reduce the size, and the research plant under construction will have a capacity of some 100 - 150 $kg\ hr^{-1}$. Much careful thought has gone into minimising any way in which the criteria above might be compromised - particularly point (vi) above - and we are satisfied that the plant will achieve its objectives. The tube bundles on a full scale plant will be assembled in a modular fashion. The research plant design is based on half a full size module, and care has been taken to ensure that the hydrodynamics of the liquor distribution from tube to tube on the research and full scale plants are similar.

The engineering design of a number of items within the research plant has presented interesting challenges, and a number of these are detailed below.

4.1 The Vapour Compressor

Two suppliers of suitable centrifugal compressors for the research plant were identified. Unfortunately one manufacturer withdrew his quotation on technical grounds at a late stage. The other indicated that the cost would be very much higher than originally quoted, and that delivery times would be extended. Both manufacturers have assured Courtaulds that the problems lie in designing a small unit for the research plant, and the production of larger machines (for a full scale plant) represents much less of a problem.

The use of a fan or single stage centrifugal compressor is an important feature of this process, and difficulties in this area cannot be lightly dismissed. Accordingly it has been decided that the use of a centrifugal compressor at the research stage must remain a prime objective, even though the cost may be high and the delivery long. So as a temporary expedient a rotary lobe compressor will be employed initially in order to avoid holding up the project. If the rest of the research plant is shown to operate satisfactorily, Courtaulds will then have the option of replacing the compressor with a centrifugal unit.

The rotary lobe compressor will be constructed in cast iron with the process contact surfaces protected by a hard nickel plating.

The centrifugal compressor is specified to have a steel casing with an epoxy protective coating. The impeller, which revolves at 12,000 rpm in this small unit, will be machined from solid titanium.

4.2 The Containment Vessel

A large vacuum vessel is required to contain the tube bundles for the evaporator. It has been decided to build this vessel in concrete for both research and full scale plants. Courtaulds have already had some success in the construction of such concrete vacuum vessels. The inside of the vessel must be protected from attack by the acid viscose process liquors, and a corrosion resistant coating will be used. If there is any porosity of the concrete, full atmospheric pressure could be developed across sections of this lining, so it is important that there be good adhesion between the coating and the concrete substrate.

A number of protective linings have been identified as being potentially suitable, and a series of laboratory experiments have been set up to investigate their performance. The method employed was to coat a 150 mm cube of porous concrete with the lining to be tested, immerse the cube in process liquor and then attempt to blow the coating off by application of pneumatic pressure to the interior of the cube. The equipment used for this test is shown in Figure 5. Additional tests are being carried out in which the mechanical force required to detach the coating is measured.

First results indicate that no problems should be encountered in specifying a protective lining which will withstand the process conditions.

4.3 The Tube Bundle

There are three major components of the tube bundle. The frame is to be constructed of glass fibre reinforced resin in pultruded sections. The tube plates are made from glass fibre reinforced furane resin, and will contain stainless steel ribs to give strength and dimensional stability. The thin walled tubes, manufactured from low density cross-linkable polyethylene, have been extruded and are ready for assembly into the bundle.

A good deal of attention has been paid to the method of fastening the tubes into the tube plate, and also to ways of taking up thermal expansion of the tubes and maintaining them under slight tension. It is clear that all these arrangements must be simple and cheap to construct as well as being able to withstand the hostile environment within the evaporator.

Figure 6 illustrates the concrete containment vessel and the positioning of the tube bundle within it, along with some of the more important dimensions.

5. Future Work

The next few months will see completion of the research plant, followed by its start-up and operation. It is planned to run the plant for some six months initially in order to assess its long term running potential.

6. Conclusions

i) Courtaulds is developing a low energy evaporator for the removal of water from viscose process liquors. The process employs mechanical vapour recompression, and uses thin walled plastic tubes as a heat transfer surface. This relatively cheap medium means that surface areas can be large, with low temperature driving forces across them. This leads to a thermodynamically efficient process, and mechanically simple compressors.

ii) The research programme was designed to last 34 months and proceed in three phases. These were: laboratory investigations into the properties of the process and constructional materials, together with the building of a bench scale evaporator; the building of a research plant at a production site, and a detailed technical, economic and commercial assessment of the process. At the present time the first phase has been completed, and the second is being executed.

iii) A comprehensive set of relevant physical properties of viscose process liquors has been obtained from laboratory work. Measurements have been made of viscosities, densities, boiling points, thermal conductivities and specific heats.

iv) The mechanical and physical properties of thin walled cross-linkable polyethylene tubes have been examined. Rates of cross-linking, wettability and resistance to fouling have been determined, and measurements have been made of thermal conductivity and coefficient of thermal expansion. The long term dimensional stability and resistance to creep rupture have been studied.

v) A bench scale vapour recompression evaporator has been constructed and operated. This demonstrated the principles of mechanical vapour recompression as applied to viscose process liquors. Heat transfer coefficients of 800 - 1,000 $Wm^{-2}\,{}^{o}K^{-1}$ were measured.

vi) A research plant with a capacity of 100 - 150 kg hr^{-1} of water evaporation has been designed and is now being built. This plant has to satisfy a number of criteria to ensure that results from it will be useful in the design of a full scale unit. The design of the vapour compressor, the containment vessel and the tube bundle have all presented interesting engineering challenges, and a strategy to meet these has been devised.

vii) The results of the research programme so far are encouraging, and the next steps involving the building and operation of a research plant are being vigorously pursued.

DISCUSSION

Question - Dr. Spendel, IKE : How much energy can be recovered ?

Answer In 1980 it was calculated that if the results of this work were successfully applied to 50 per cent of the viscose production in the EEC (excluding Courtaulds Group), the savings would amount to 1.7×10^6 GJ/year (about 5×10^5 MWh/year).

Question - Dr. Spendel, IKE : Is it useful to remove the non-condensible gases (e.g. by absorption) in order to improve the condensation heat transfer in the tubes ?

Answer The removal of non-condensibles will undoubtably improve the condensation heat transfer in the tubes.
Degassed and undegassed feedstocks will be used to determine the effect of non-condensibles on condensation heat transfer in our research plant. We will know whether it will be economical to provide degassing facilities in a commercial plant. At present, we suspect that it will not be so.

Question - Mr. Grehier, IFP : What is the improvement in heat transfer coefficient you obtain by the 'cross-linking' treatment of the polyethylene plastic ? Is this cross-linking treatment specific to Courtaulds ?

Answer We have never made a direct comparison of heat transfer coefficients because it seemed pointless to measure these coefficients on surfaces which were clearly non-wettable. However, to judge by our qualitative observations, the coverage of the plastic surface by the process liquors increases from about 30 per cent to over 80 per cent when the surfaces are made wettable. So it would seem reasonable to expect the coefficient (on the liquor side) to increase by a factor of perhaps 2-3.
The chemical cross-linking treatment is not specific to Courtaulds, although the use of viscose process liquors as a catalytic agent is.
Courtaulds do have processes for making many plastics wettable, but they are not required for this project.

Question - Dr. Pilavachi, CEC : Do you have any potential buyers of this process outside your company ?

Answer We have not yet approached any viscose producing companies with a view to selling this equipment. However, Courtaulds has an established precedent in selling fibre-producing plants world-wide through its engineering company, Courtauld Engineering Limited.

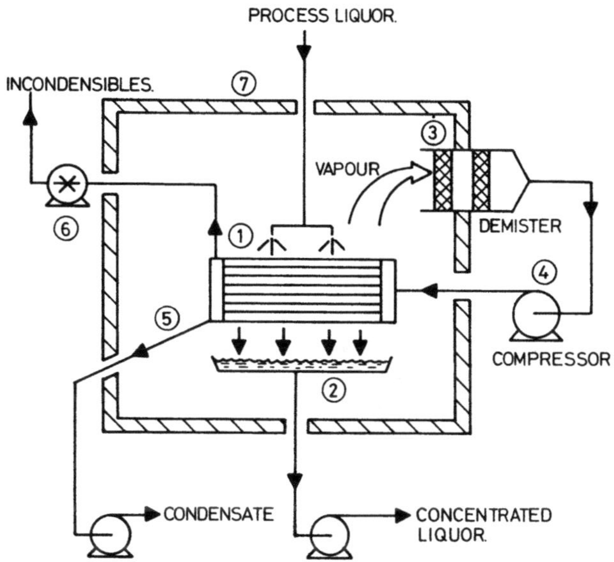

FIGURE 1.

ESSENTIAL FEATURES OF A MECHANICAL VAPOUR RECOMPRESSION EVAPORATOR FOR VISCOSE PROCESS LIQUORS.

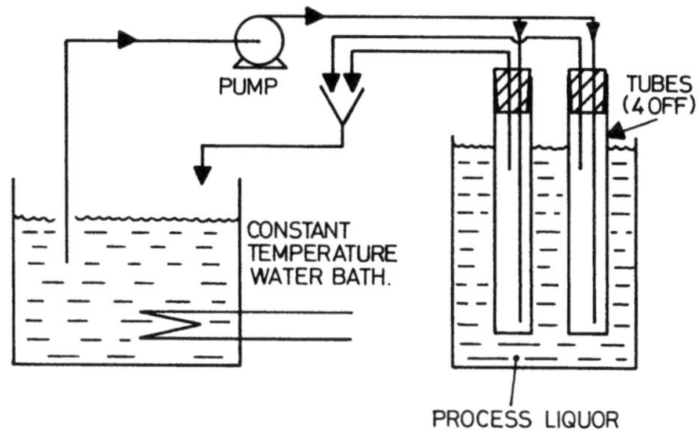

FIGURE 2.

SCHEMATIC DIAGRAM OF APPARATUS USED TO CHECK FOULING OF EVAPORATOR TUBES.

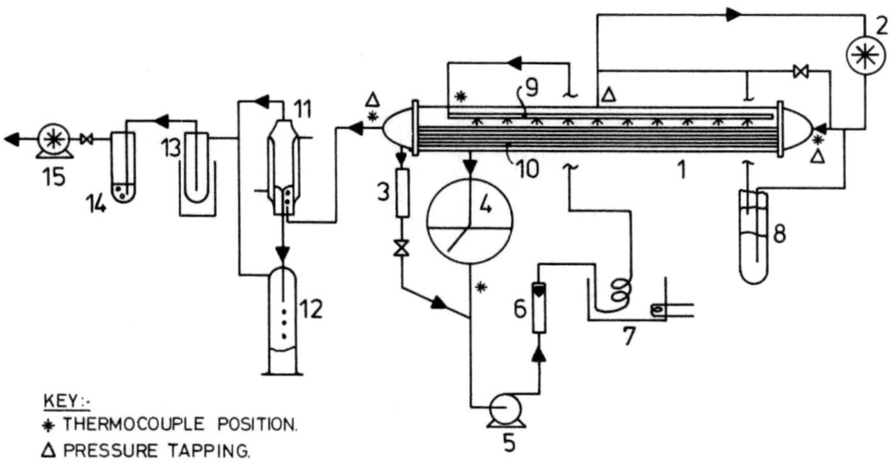

FIGURE 4

DIAGRAM OF BENCH SCALE EVAPORATOR

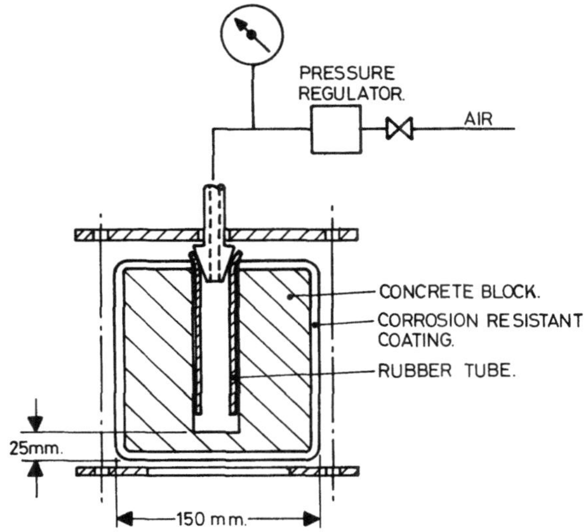

FIGURE 5

APPARATUS TO TEST ADHESION OF CORROSION RESISTANT LINING TO CONCRETE.

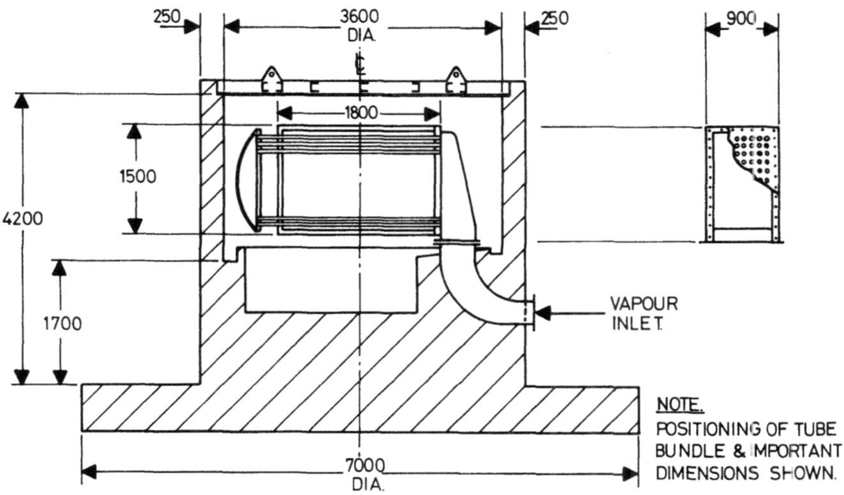

FIGURE 6

SKETCH OF CONCRETE VACUUM VESSEL FOR RESEARCH PLANT.

ALUMINIUM MELTING FURNACE HEAT REGENERATION PROJECT

Author	:	DONALD McCALLUM
Contract No.	:	EEB.1.151.80.UK
Duration	:	1981/82
Contractor	:	James Howden & Co. Ltd.
Address	:	195 Scotland Street Glasgow G5 8PJ United Kingdom

SUMMARY

The aim of the project is to extend the range of the Howden rotary regenerative preheater into the high temperatures associated with metal remelting furnaces. The initial objective is to design a metal regenerative air preheater suitable for use on an aluminium melting furnace with the high temperature gas diluted to a maximum of 750°C.

The Research programme has proposed the installation of a single trial preheater operating on two aluminium remelting furnaces at Alcan Plate Ltd., Birmingham. The programme at present has covered the measurement of gas flows, temperatures, gas analysis, dust burdens and flow patterns.

Other than the technical problems associated with the process the main difficulty has been in finding a user willing to instal the first system into their furnaces. Our original contract (No.386-78EEK) involved Alcoa Manufacturing (G.B.) Limited, whose furnaces are now scrapped and our present contract has been delayed due to the cut-back in demand for aluminium plate for the aircraft industry.

1. INTRODUCTION

Sixty years ago Fredrick Ljungstrom, a Swedish Engineers invented and patented the Rotary Regenerative Air Preheater and James Howden the first of his major licencees have since successfully supplied over 3000 units to the Power and Process Industries.

The standard mild steel design supplied by James Howden is limited to a gas inlet temperature of 450°C and an air preheat of approximately 400°C. This present research contract is involved in the extension of the normal design using carefully selected materials and a novel form of construction to extend the temperature range to a gas inlet temperature of 750°C and an air preheat of 600°C.

The gas temperature being limited to 750°C by the suitability of present high temperature corrosion resistant materials. The air outlet temperature being limited to 600°C to suit available oil fired burners suitable for use with this degree of air preheat. This level of preheated air allowed the use of Burners without the additional problem of water cooling.

The original project (No. 386 - 78EEK) was formally started in 1979 however before the hardware was put on order the original user cancelled and closed the furnace. It was decided with E.E.C. approval to investigate alternative users with whom a similar scheme could be tested.

Fortunately an alternative site was identified and the present contract agreed with E.E.C. however the final commitment by the end user allowing the hardware to be ordered has not materialised. Complete duplication of all stages of our original contract has been necessary and a very complete design study had been submitted to our customer who have now accepted our proposals technically but at this moment are delaying their final commitment due to large financial losses in other divisions of their Company. Alternative financing proposals have been offered and we await their acceptance.

2. DESCRIPTION OF APPARATUS

Furnace Details

The furnaces to which air preheat has to be applied are the melting sections of a handed pair of Salem - Brosius aluminium melting and holding furnaces.

Each furnace has a capacity of 25 tons.

The furnaces are rectangular with a sloping roof, charging door at the high end and gases exhausting through the top of the low end. They are fired by five burners, three in the sloping roof and two in the side walls.

Operation

The normal approximate operational cycle is 3½ to 4 hours duration. The charge varies from large ingot to comparatively thin scrap. The furnaces are charged by a skip on the end of a hydraulic ram through the end door. The skip is in turn loaded by fork lift truck using the space in front of the door and between the furnaces. This space is also used for storage of charge before loading.

Normally all burners are off when charging - stirring - or slagging through the main door. Both furnaces are fixed at their maximum firing rate after charging until the charge is within 10 ^{o}C of the bath temperature of set point.

The charge is frequently held in the melting furnace for appreciable periods after melting is completed.

Layout

The layout of the installation and the design of the control system is based on the premise of a minimum amount of change to existing equipment and operation. It must be possible to immediately revert to the original method of operation.

A typical layout arrangement for a single furnace preheater installation is shown in figure 1 and the control system layout for the two furnace arrangement is shown in figure 2.

Air Preheater

In regenerative heat exchangers, the heat transfer surface is alternatively heated by the flue gases passing through it and cooled by the air passages at different times so that unlike the recuperative heat exchanger where the heat flows through the passage walls from the flue gases to the air, the heat is absorbed by the regenerative mass from the hot flue gases and regenerative mass rotates as in the Howden Air Preheater, the process is continuous.

In the Howden Air Preheater, flue gas flows through one side of the generator and through the other side flows the incoming air prior to entering the furnace. The regenerator is slowly revolved so that the heating elements pass alternatively through the stream of hot flue gases and through the stream of cold air. A part of the heat in the flue gas side is transferred to the elements when they pass through the flue gas side and this heat is transferred to the air when the elements pass through the air side, so cooling the elements. Thus the heat in the flue gases is partly recovered and returned to the furnace.

The two streams, flue gas and air, which flow through diametrically opposite segments of the rotor, are separated from each other by a small blanking section with sealing plates to form a division between them. The two streams flow in opposite direction i.e. in contra-flow gas down air up.

Heating Surface Elements

The elements will be made from stainless steel in a corrugated undulated form. The corrugated sheets to form a large number of open flow passages and the undulated sheets to provide a turbulence to the flow of air and gas.

Vanadium measured in the gas analysis has made it necessary to split the element length into two equal depth of elements.

Cold End	600 mm 9.5 c.u.	Stainless steel
	0.545 mm thick grade	Type 321 S15
Hot End	600 mm 9.5 c.u.	Stainless steel
	0.8 mm thick grade	Type 253 MA

It is also proposed to supply two hot end sectors of elements in Inconel Alloy 601 which would be installed from the beginning to assess the improved life of this more expensive metal.

The 9.5 corrugated undulated element form is a wide pitched element which is normally used in coal fired power stations, and has been selected in this instance to cater for the possible dust and surface oxidisation.

Element Containers

These are also of substantial stainless steel of robust welded construction containing the heating surface elements. These containers or packs are designed for easy handling into and out of the preheater.

Rotor

Comprising a lagged and air cooled central steel shaft of fabricated construction with bolted stainless steel radial division plates arranged to give twelve separate compartments or sectors. Each sector would contain two element containers, one hot and one cold container, bolted to the peripheral edge. A grid to be welded at top and bottom to provide additional baffle seals, the bottom grid also acting as a support system for cold end containers. The top, bottom and outer edges of radial plates would have bolted adjustable seals fitted as described later.

Casing

The rotor is contained in an internally insulated mild steel casing of suitably stiffened panels. Each duct connection is flanged for bolting. Between the gas and air openings at each end of the housing are sections forming the sealing surfaces for radial and circumferential seals attached to the rotor. The insulation will be of the ceramic fibre type suitably supported and clad in thin stainless steel panels. For removal of the element packs, a door is provided on one panel which allows the containers to be loaded or unloaded axially.

Seals Between Gas and Air Sides

Fabricated Inconel construction developed for the particular application and comprises:-

(1) Radial sealing strips at each end of the rotor and attached to the radial division plates of the rotor but capable of adjustment and forming a seal with the sections of casing and closing plates located between the gas and air openings.

(2) Circumferential sealing strips at each end of the rotor at the outer and inner periphery attached to the rotor but capable of adjustment and forming a seal with the machined surfaces on the casing and closing plates.

(3) Axial sealing strip attached to the rotor positioned axially along the cylindrical shell of the rotor at locations corresponding to the ends of the rotor radial division plates and forming a seal with the curved surfaces on the axial seal plates attached to the inside of the Preheater end pillars.

(4) Hub sealing strips at each end of the rotor, attached to the end closing plates of the casing and forming a seal with the periphery of the rotor shaft. The sealing strips are adjustable.

Sootblowers

Single nozzle traversing sootblowers are supplied for the Preheater, one mounted in the gas inlet ducting blowing with the gas flow and one in the gas outlet ducting blowing against the gas flow.

Unlike the much published 'Heat Wheels' which have been developed from the Ljungstrom design the Howden Regenerative Preheater is of robust construction to suit dirty and corrosive environments found in power stations and process industries. These devices are capable of cleaning the heating surfaces 'on load' and the elements and containers are suitably constructed to absorb the high energy cleaning medium.

Maintenance

All items in the preheater likely to require replacement due to hazardous service have been designed for each replacement.

3. SITE TESTS

Test 1

Sample element packs of stainless steel type 321 were installed into the flue gas ducts of one of the furnaces. Unfortunately the gas temperature could not be controlled and peaked to 900°C.

Weight tests and dust tests were made on two packs the first pack being removed after 7 weeks, replaced and given a further 16 weeks before finally being removed.

During this second period of the test a plain Incoloy sheet was added to the test pack.

Weight Loss Report

	Pack 2 (28 weeks)	Pack 1 (7 weeks)	Pack 1 (16 weeks)
Weight before test	3271 gm	3273.5 gm	2909 gm
Weight after test	2732 gm	3023.5 gm	2834 gm
Weight loss	539 gm	250 gm	75 gm
% Weight loss	16.5%	7.6%	2.6%
Weight of Incoloy Sheet before test			163.5gm
Weight of Incoloy Sheet after test			163.5gm
Weight loss			0 gm

Analyses of Dust Samples (After 28 weeks)

Carbon, C	1.80	0.15	8.34
Silica, SiO2	19.82	7.50	10.80
Manganese Oxide, Mn3O4	2.02	1.00	1.06
Alumina, Al2O3	10.32	3.06	8.20
Ferric Oxide Fe2O3	21.46	62.61	22.45
Chromic Oxide, Cr2O3	1.93	13.37	3.02
Nickel Oxide, NiO	1.93	4.96	2.65
Sulphate, SO4	18.83	3.72	21.99
Vanadium Pentoxide V2O5	1.50	1.18	1.59

These results indicate oxidation (scaling) such as would be expected with temperatures up to 900°C the most significant measurements was the presence of vanadium.

High temperature corrosion is resisted by stainless steels with the formation of a thin oxide layer over the surfaces. Vanadium pentoxide attacks this oxide layer and it is generally accepted that this becomes significant above the melting point of the compound (660°C). Accordingly the element packs will be split into two with the hot end material being supplied in stainless steel grade 253 MA which has a resistance to this type of attack.

If any attack does take place this will only be expected at the extreme edge of the element material. With the construction of Howden preheater it is possible to increase the life of these hot end elements by turning the element packs prior to complete deterioration of the corroded edges. The regenerative principle is not effected by the thickness of the material on the oxide layer on the surface

Test 2

A research division of British Steel was contracted to complete a series of extensive tests which included a lengthy record of furnace operation, temperatures, flows, gas analysis, oil consumption and aluminium quality and quantity. The results of these tests were initially used for the design of the control system and the assessment of the part load performance of the preheater and furnaces.

A further series of tests will be completed after the installation is commissioned which will be used for a 'before and after' comparison.

Figure 3 shows the temperature pattern of one furnace during one typical cast.

4. SHOP TESTS

As the site tests shows the presence of vanadium in the flue gas it was deemed necessary to alter the matrix materials and element split. Two alternative materials were selected, sample pieces purchased and the formability tested to ensure that correct element profiles would be obtained. Both new materials gave good results.

5. CONCLUSIONS

The design of the preheater and controls have been assessed in detail by Alcan Plate Limited, who have concluded that the installation as proposed is most suitable for their furnaces. Comparisons were made of recuperators which could only achieve 400°C air preheat. Finance for completion of project is restricted however it is hoped that an order will be forthcoming in the near future.

DISCUSSION

Question - Chairman : What leakage do you expect on this type of unit ?

Answer We do not expect the leakage to be more than 5 per cent on this project.

Question - Mr. Galant, Bertin : What is the investment cost of the air preheater and what are the flow rates ?

Answer The preheater cost is approximatively £ 100,000 (180000 ECU). However, the project being a retrofit involving two furnaces with necessary new burners and controls, the total investment is of the order of £ 400,000 (700000 ECU). Even so, the payback period on present furnace operation is expected to be under two years. The payback period on new installations would therefore be well under one year. Air flows for both furnaces are 16,800 kg/h.

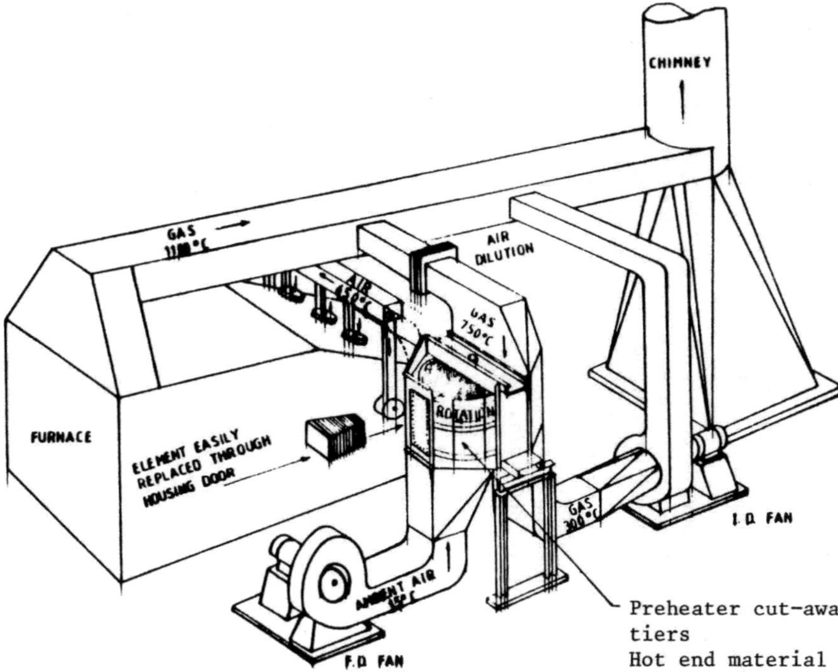

Preheater cut-away to show matrix supplied in two tiers
Hot end material selected for H.T. corrosion
Cold end material selected to suit combustion gas dew point.

Figure 1 TYPICAL ARRANGEMENT OF HOWDEN HIGH TEMPERATURE REGENERATIVE PREHEATER ON AN ALUMINIUM MELTING FURNACE (EXPANSION JOINTS & CONTROL CAMPERS NOT SHOWN)

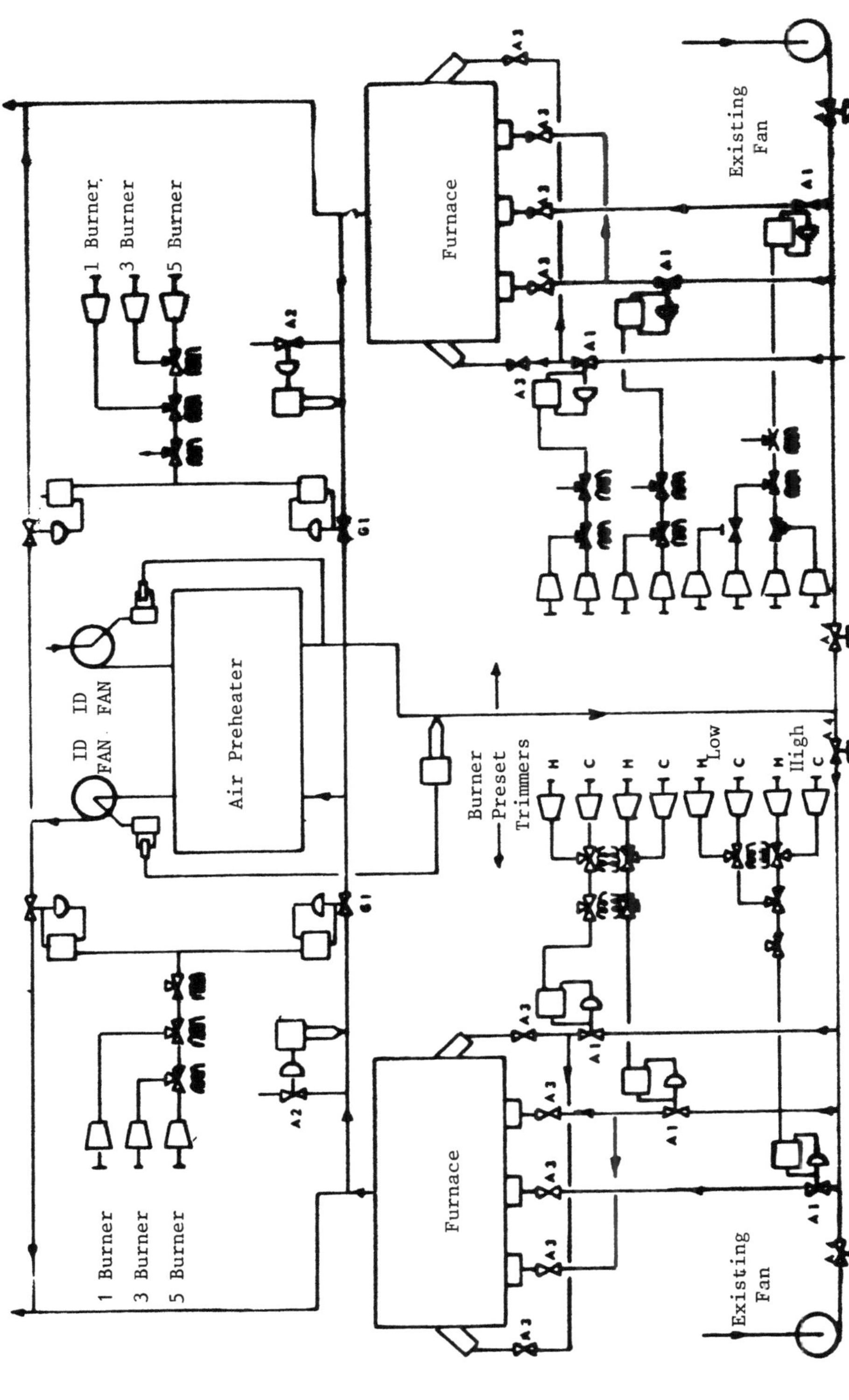

Figure 2

CONTROL SYSTEM

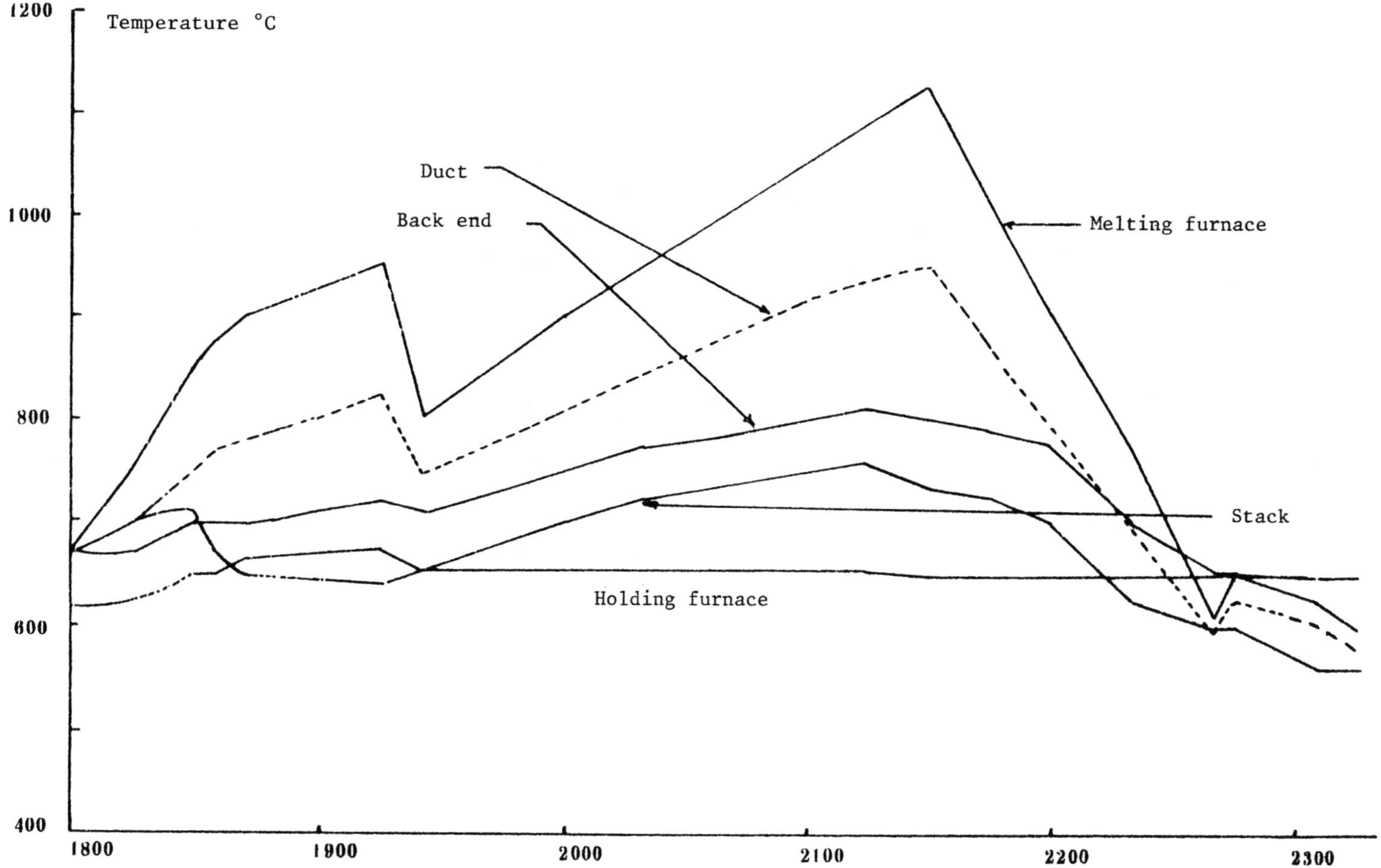

TYPICAL EXTRACTION TEMPERATURES

Figure 3

DESIGN OF WASTE HEAT BOILERS FOR THE RECOVERY OF ENERGY FROM ARC FURNACE WASTE GASES

Authors	S. BRAMFOOT, J. DIXON, J. R. MARTIN A. D. ROBERTSON
Contract Number	EE-B-2-160-UK
Duration	24 months 1 MAY 1981 - 30 APRIL 1983
Head of Project	J. Dixon
Contractor	British Steel Corporation, Sheffield Laboratories
Address	British Steel Corporation, Sheffield Laboratories, Swinden House, Moorgate, ROTHERHAM GB South Yorkshire. S60 3AR

Summary

A literature survey has revealed a wide area of general interest in methods of waste heat recovery from the electric arc furnace although quantitative information tends to be conflicting. An analysis of test results from BSC arc furnaces has been used to provide a typical pattern of waste heat rejection. This data provided a waste heat rejection of 200 kWh/tonne for a 170 tonne furnace. The waste heat boiler application is discussed in relation to the scrap preheating technique currently under consideration in BSC. Steam usage by the Works is examined and typical steam requirements are identified as a basis for evaluation of the boiler design. The current and future work programme is outlined as a means of obtaining a reliable economic assessment of the application.

1.1 Introduction

At present the electric arc furnace accounts for 30% of the steelmaking capacity in Britain. The cost of electrical energy is therefore a noticeable proportion of steel production by the electric arc route. It has been estimated that the energy rejected in the waste gas could amount to over 150 kWh/tonne of steel for a large 170 tonne furnace. The research project is intended to recover this heat from the waste gases using a waste heat boiler and will consist of two phases. The first phase is a design study to establish the feasibility and approximate cost of a boiler application and is the subject of the present work. If this proves successful a design and construction of a prototype boiler will be developed with performance tests at a melting shop in BSC. The paper describes the progress so far and outlines the future course of the work.

1.2 Plant Description

The electric arc furnace operates as a batch process with the volume, temperature and composition of the waste gases varying throughout the cycle. The furnace is top-charged with cold scrap which is melted, refined by oxygen injection and tapped into the ladle over a cycle period of 2.5 to 3 hours. Emissions of hot combustible gas (including CO and H_2), slag particles and ferric oxide fume are treated in a multi-purpose extraction system (See Figure 1). During melting and refining the waste gas temperatures may reach 1850°C. The gases are extracted from the furnace through a water-cooled elbow fitted to the roof. The gases are burned in a suitable combustion duct and the hot combustion gases are cooled by dilution air and/or by water sprays to a safe working temperature before cleaning. Air for combustion is infiltrated at a slice gap between the roof elbow and the combustion duct to facilitate roof swing and furnace tilt. During the charging and tapping stages the furnace is disconnected from the direct extraction system and fume emission is collected by hoods in the shop roof and ducted to a bag filter plant.

1.3 Waste Gas Conditions

A literature survey revealed little detailed or useful information on waste heat rejection from the arc furnace and such as there was tended to be conflicting. The energy input may be defined as electrical and chemical, the latter covering both bath oxidation (primarily carbon) and the combustion of electrode graphite. Three publications examined on this basis provided the following energy split.

Reference	Energy Split % electrical	 chemical	Energy transferred to steel & slag %
1	63.5	36.5	59
2	80 - 90	10 - 20	50 - 70
3	60 - 80	20 - 40	52 - 61

A more consistent picture was obtained by expressing the information as the percentage of the total energy rejection which is transferred to the bath (the fourth column in the table). The balance of this energy would be dissipated in various losses including heat lost in the gases. The data quoted in the above references[1,2,3]. for the energy in the furnace gases were 15.5%, 5-9% and 19-20% of the total energy supplied. These correspond to heat rejections of 100 kWh/tonne 30-60 kWh/t and 130 kWh/t. In view of these variations it was considered more appropriate to use data collected from plant information available within BSC, including the results of furnace trials. The basic data for the BSC furnaces is given in Table 1.

Depending on the fume separation plant the gas extraction facilities and their operation can vary enormously. Any waste heat boiler would be fitted into this area (see Figure 1). Furnaces E, F, T & C and the stainless furnace have their direct systems connected into the roof ducting for cleaning in a bag filter plant. The remaining furnaces have separate direct extraction facilities with electrostatic precipitators for gas cleaning. In the former case, the air/gas ratio used is high to maintain a low gas temperature at the bag plant. The combustion gas temperature for the separate system is maintained at below 1250°C to protect the linings from molten slag attack and further cooling is provided by water sprays to a safe precipitator temperature. The analysis of the plant data showed significant variations in waste gas conditions both during the furnace cycle and between furnaces. An example of these variations is shown in Figs 2 and 3 for E furnace. From an overall consideration of the data a typical pattern of waste heat rejection for a 170 tonnes furnace was obtained and this is shown in Fig 4, giving a mean of 200 kWh/t.

This data provides a heat rejection of 9 MW for 50 minutes 20 MW for 60 minutes and 11 MW for 40 minutes. Based on the working year, the total energy loss would be 70 GWh/year, equivalent to an electrical energy cost of £2 x 10^6 per year. Recovery of this heat to replace fuel oil for steam raising would provide a cost saving of about £0.5 x 10^6 per year. Applied to only half of the arc furnace production in the

European Community the recovered energy cost would amount to £20 x 10^6 per year. The real savings would, of course, depend on the capital and operational costs to be offset against this return.

1.4 The Approach to Waste Heat Recovery

As stated above, the waste heat boiler would fit into the direct extraction system between the slice gap and the transfer ducting leading to the cleaning plant. However, reference must first be made to scrap preheating, an attractive method of heat recovery whereby reject heat is returned to the furnace. Essentially the furnace charge is heated by burned gas either in the charging basket or in a rotating kiln and both methods are being considered for certain arc furnaces in BSC and must therefore be considered in relation to the project. Work done so far has shown the process to be limited by the temperatures which the scrap containers can tolerate and there is significant residual heat in the exit gas flow. Also a scrap preheater may be too costly to install in an existing melting shop, a constraint which could also apply to the waste heat boiler. Nevertheless, a waste heat boiler may be considered in series with a scrap preheater or as an alternative process where installation costs determine such a course.

1.5 Works' Demand for Steam

A survey of the works' steam usage showed a considerable variation in the application of steam and the quality required. For example, melting shops have on-site requirements to seal electrode ports against fume and gas emission and to operate ejectors in vacuum degassing units. In the mill areas steam is used for oil atomisation in reheating furnace practice and also for jacket heating of units such as pickling tanks and fuel oil lines. In all areas steam is used for shop and office heating. Three examples taken from the sites surveyed are given below.

Works 1

The boiler plant supplies hot water initially at 150°C and 11 bar to provide space and amenity heating in two melting shops (3 arc furnaces in total) and a spring shop. In winter the demand is stated as 500,000 kg/h which in summer drops to about 182,000 kg/h. The system is based on a pressurised nitrogen balance operating as a closed loop.

Works 2

The plant consists of a three-furnace melting shop and several mills spread over a wide area. During March, 1982 the steam demand varied from about 15,000 to 30,000 kg/h over seven days, whilst during the working week of five days the variation was 23,000 to 30,000 kg/h. At the weekend steam is required for space heating and to maintain temperature at certain areas such as oil supply lines. At this works the steam is supplied at 204°C and 11 bar. Under the circumstances the demand is not as variable as might be expected.

Works 3

At this works steam is required for process purposes, hot water is required for space heating and high grade steam for power generation. Steam is used at 45 bar pressure and 425°C to generate electrical power. Steam is recovered at 14 bar after the first stage of the turbine. Part is used directly as process steam and part is sold to adjacent external customers. The waste steam is condensed in calorifiers to provide hot water for space heating. When space heating demands are high, the low pressure supply to the calorifier may be supplemented by higher pressure process steam.

It is now clear that any assessment of waste heat recovery from arc furnace waste gases can only be done on a Works to Works basis. However, for the purposes of a general approach, there are four basic options for steam usage:-

(i) the generation of hot water.

(ii) the generation of low pressure steam (5.5 bar and 160°C) for process applications relatively close to the boiler plant.

(iii) the generation of steam at higher pressures (15 bar and 200°C) for process applications remote from the boiler unit.

(iv) applications (ii) and (iii) refer to steam from a waste heat boiler to supplement an existing works' supply line. Alternatively, high grade steam at say 45 bar and 425°C could be produced to generate electricity using a steam turbine.

1.6 Boiler Design

The next stage in the course of the project will be to determine specific details of the cost and design of a waste heat boiler. Clearly technical and economic advice, and possibly co-operation, will be required from professional boiler manufacturers. Accordingly, contacts are being made and discussions are in progress based on the results of the foregoing surveys and studies. A work programme has been determined and is summarised overleaf.

Boiler Performance

The first priority is to determine the type, size, space and performance of the boiler unit and this should arise from the discussions with the boiler manufacturers. Boiler performance will depend critically on the heat transfer conditions that apply for dirty gas flows. Plant trials are therefore planned for these conditions to obtain technical and cost information in terms of performance and maintenance.

Engineering Aspects

From the variations in both waste heat available and steam usage it must be decided whether to consider several boiler designs for the various circumstances or to concentrate on a specific design for one works. In either case three items require detailed study:-

- costs and savings relevant to the basket preheater in relation to the boiler application

- the problems and costs of different melting shop layouts in relation to boiler installation

- a cost study of boiler design in relation to the various extraction plant options.

The engineering considerations will be ongoing during the progress of the boiler performance work.

Initial discussions with boiler manufacturers have suggested that the combustion chamber would not be a suitable location for a boiler. A more effective site would therefore be a unit in parallel with or combined with the ductwork and/or conditioning chamber. It was stated also, that a steam storage facility should be considered primarily for steam used to feed the Work's supply main. The storage unit would take the form of a hot water accumulator.

2 Conclusions

Preliminary considerations suggested that significant amounts of heat rejected during electric arc steelmaking might be recovered by raising steam in a boiler unit. A subsequent literature survey revealed a wide interest but little specific information on the subject including waste heat flow. Therefore, the results of plant trials done on BSC furnaces were analysed and a typical pattern of waste heat rejection was obtained. On this basis the waste heat available for recovery is 200 kWh/tonne of steel for a 170 tonne capacity furnace. This is equivalent to an annual loss of 70 GWh and represents a potential annual saving of £0.5 x 10^6 per furnace if used to replace fuel oil for steam raising.

Scrap preheating is also being assessed for certain BSC arc furnaces and the waste heat boiler must be considered within this context. However, scrap preheating is subject to limitations particularly regarding the tolerable preheat temperature. There is, therefore, scope for a waste heat boiler application either as an alternative method or in combination with a scrap preheater.

A survey of steam demand within BSC sites with arc furnaces has revealed that specific uses for steam vary from Works to Works. In general, the Works' boiler plant may be asked to provide hot water or steam for space heating, process steam or high grade steam for power generation. On this basis and excluding hot water, three types of steam have been identified for discussions with boiler manufacturers. A programme of work has been formulated on the basis of the foregoing work and priorities have been identified in order to implement the work.

3 References

1. Blank, J.R., Garbutt, R., 'Energy Conscious Electric Arc Furnace Steelmaking'.

2. Pautz, J, 'Recovery of Energy from Arc Furnaces, Electrowarme International; Vol 37, No. B2, pp 86-92, April, 1979.

3. Fernadez Lopez M, et al., 'Uses of Arc Furnace Fume Energy Cenim Revista de Metalurgia 15 (1) 1979, Jan/Feb pp 43-56.

DISCUSSION

Considerable interest was shown in this project, particularly in the pattern of supply of and demand for steam if a waste heat boiler is used as a heat recovery device on arc furnaces.

Question - Dr. Pilavachi, CEC : How can an intermittent steam supply be fitted in a plant ? If there is no simultaneous steam demand, there may be some steam blown off to atmosphere. The economics of the head recovery system would then be less interesting.

Answer It is our view that some form of steam storage in, say, a hot water accumulator will be necessary to cope with fluctuations in steam supply and demand. This is one area for discussion with the boiler manufacturers. Obviously our objective would be to avoid a blow-off situation if at all possible.

Question - Dr. Thornton, Courtaulds : Do you envisage any technical advances in this field of heat recovery, or are you concerned solely with the applications of existing technology ?

Answer It is difficult at this stage to foresee any specific technical advances arising from our project, and we should rely at the outset on existing technology gained from boiler users and manufacturers. However, the application is new, at least in BSC, and the adoption of new techniques and developments may well arise.

Question - Mr. Rojey, IFP : Does steam represent a significant part of the energy consumed in the steel industry, and is there really a demand for the steam produced by a waste heat boiler ?

Answer The steam usage varies enormously from Works to Works and we have learned from our surveys to consider each Works on its own merits. If our further work shows a boiler and turbine arrangement to generate electrical energy from arc furnace waste heat to be a feasible and convenient application, then the demand could be more universal.

Question - Mr. Galant, Bertin : Could you give exit temperature levels, together with available waste heat power, as a function of time ? They are both needed to design recuperative systems.

Answer This is perfectly true and temperature measurements, as well as heat transfer measurements, will be part of our forthcoming furnace trials. At the present stage however, we have been concerned only with the quantity of the heat available, irrespective of the thermal drive which in any case can be adjusted by variation of the air infiltration into the combustion chamber.

Comment - Mr. Galant, Bertin : We have just carried out deposit tests on tubes which show that heat transfer rates drop essentially because of the conductive barrier they cause. It is then necessary to know deposit rates in advance and to predict new lower heat transfer rates.

BSC : We are prepared for a reduction in heat transfer with build-up of oxide and for slag on the tube surfaces. Our probe tests on the furnace will be designed to study this factor and hence obtain a realistic picture of boiler efficiency to be expected. In this respect, therefore, we should find your own deposition work particularly interesting.

WORKS	FURNACE	NOMINAL CAPACITY (T)	SHELL DIAMETER (M)	RATED POWER (MVA)	OXY-FUEL BURNERS
ROTHERHAM	E	170	7·3	80	-
	F			80	-
	K			56	-
	N			64	-
	T			64	-
STOCKSBRIDGE	B	120	6·1	45	-
	C	150	6·7	35	3
	D	150	6·7	35	-
TINSLEY PARK	Y	135	6·7	60	-
	Z	135	6·7	60	-
	STAINLESS	130	6·7	60	-
RIVER DON	X	100	6·4	35	1

TABLE 1

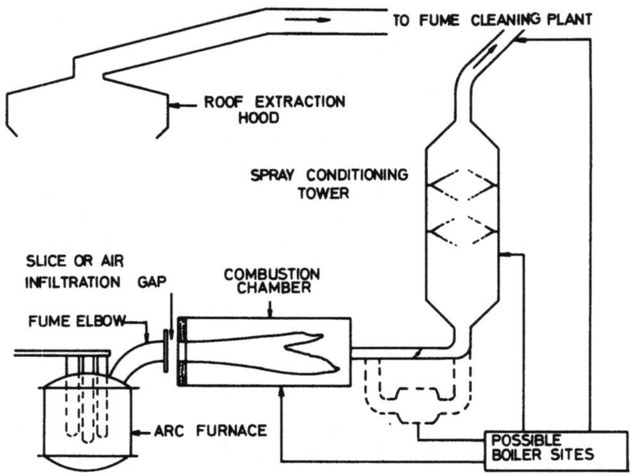

GENERAL LAYOUT OF ELECTRIC ARC FURNACE AND ASSOCIATED WASTE GAS DUCTING.

Fig. 1

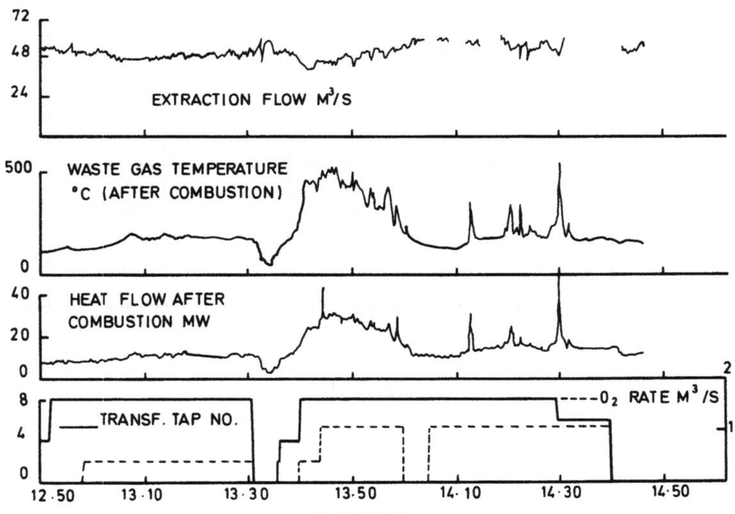

WASTE GAS CONDITIONS-'E' FURNACE, ROTHERHAM WORKS

Fig. 2

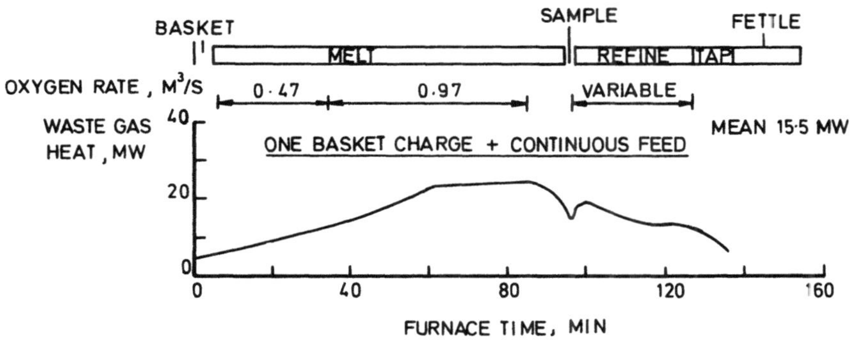

AVERAGE WASTE HEAT CYCLE-'E' FURNACE, ROTHERHAM WORKS
NOMINAL CAPACITY 170 TONNE

Fig. 3

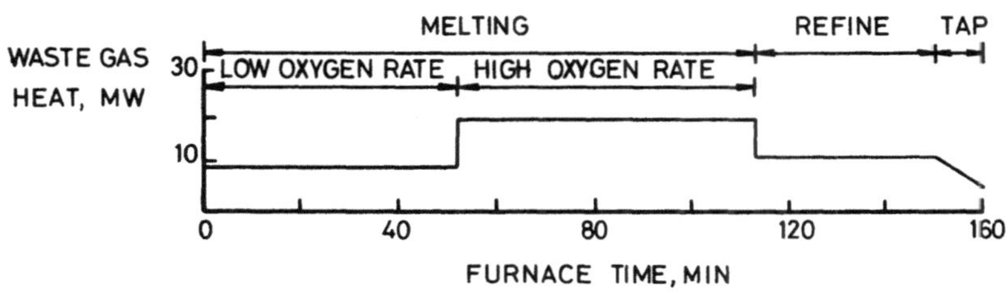

MEAN WASTE GAS HEAT CYCLE FOR A 170 TONNE CAPACITY FURNACE

Fig. 4

ENERGY RECOVERY FROM CAST IRON MELTING FURNACES

Authors	A. Calabrò,	Teksid SpA
	M. Mischiatti,	Costruzioni Meccaniche FBM SpA
	C. Antonini,	Costruzioni Meccaniche FBM SpA

Contract no.	EEB-2-162-I
Duration	1981-1982
Contractor	TEKSID S.p.A.
Address	Corso Mortara, 7 - 10149 Torino (Italy)

Summary

The aim of the research is to contribute to reduce the energy consumption in a foundry.
Experimental tests have been performed in a pilot plant installed at one of TEKSID (FIAT Group) foundries.

The pilot plant operation aims to demonstrate the technical and economical possibility of an industrial plant that recovers and converts the latent and sensible heat of the cupola emissions into electric energy by a recovery system composed of thermal and electrical conversion subsystem, considering that the Kwh produced can be directly utilized in the same foundry.

The energy saving test pilot plant has been studied and erected under CEE financing program; being its capacity stated in such a way to obtain the wanted information for the industrial plant. Because of economical reasons the test pilot plant has included the thermal subsystem only, whereas the turbine service is simulated by an aircooled condenser.

According to the author's point of view the research programme has demonstrated the possibility of the energy recovery industrial plant burning such a kind of waste cupola gases: the main results are here illustrated.

1. INTRODUCTION

The foundry industry is a source of basic components, chiefly for the mechanical industry, but also for other branches; it thus plays a very important role in the economy of the industrialized countries.

Because of the difficult energetic situation and in order to reduce the dependence on oil producers, the consumer countries will make every effort for the development of internal resources and obviously for the greatest cut in energy consumption.

The foundry is a great energy consumer: in Europe 45% of total consumption is for industry as a whole; 15% i.e. a third of industrial consumption is for the mechanical and metallurgical branches including foundries.

1.0. The cupola as a possible field of energy recovery

It is well known that the fumes produced during cast iron melting pass through the charges and reach the top opening with some energy both as sensible and as latent heat. The sensible heat is a function of the fume temperature, whereas latent heat is related to the varying, but usually high contents of carbon monoxide, CO.

The composition of emissions from the cupola varies very much dependent on the contents of different components; with reference to the quality we can say that there are no considerable differences from plant to plant.

There are three basic components: Carbon Monoxide, CO; Carbon Dioxide, CO_2; and Nitrogen, N_2. Besides there are small quantities of steam and sulphur dioxide SO_2, as well as a great quantity of solid particles that the gases carry with them.
CO is formed after the combustion of coke; generally the CO contents signify higher quantities of heat recoverable by its combustion, according to the reaction:

$$2\ CO + O_2 = 2\ CO_2 + 2 \times 67{,}6\ \text{Kcal}$$

Energy is gained and great ecological advantages are derived by suitably transforming the heat from CO combustion.

In CO combustion this very poisonous gas is transformed into the non-poisonous CO_2, and a serious cause of air pollution is avoided.

In this research the possibility of recovering the cupola fume heat contents by producing electric power is examined. We think that this way of heat recovery is profitable compared with others.
In fact, the need of electric power is very high in the foundry for several uses, such as dust suction, pouring lines motion, holding furnaces and sand mullers: the electric power produced by the heat recovery method can be used on the spot. Its use is not dependent on any environmental or time factor, because the power is required all day long throughout the year from the Pole to the Equator. Finally this

is a valuable energy counterbalancing as pay-out the greater investment required for production.

1.1. Cupola studied in this research

The cupola studied for recovering the waste heat in the form of electric power is the liningless watercooled cupola, which produces cast iron for the automotive industry. It can melt up to 40 tons/h of metal.

The fumes picked up under the charging door are purified with alkalinic water (ph 7.2 ÷ 7.8) in a wet dust collector, where dust is reduced from 10 gr/Nm3 to 100 mg/Nm3, SO2 present in the fumes is fixed almost entirely at 30 ppm.

In Table I the main characteristics of the cupola are shown with special reference to the parameters most closely concerned with energy recovery.

1.2. Recoverable heat and electric energy produced

The calculations have been performed considering a large capacity cupola with a melting rate of 40 ton/h. In this case the gaseous emissions flow rate is about 25.000 Nm3/h and according to Table I data we have assumed an average CO contents of 19%.

Taking into account the heat losses at the chimney where the fumes are cooled to 200°C, the recoverable heat in the recovery system is about $12{,}4 \cdot 10^6$ Kcal/h.

We have considered a steam Rankine cycle whose total efficiency is about 25% i.e. 3.400 Kcal/Kwh; according to these data the electric power developped by the recovery system at the alternator clamps is about 3.500 KW.

On the operation basis of 220 days/year and 16 hours/day the total energy produced per year and per ton of cast iron were calculated; the data are summarized in Table II.

1.3. Systems for energy recovery

The cupolas operating in the foundry industry are of different types: we will only draw the attention that there are cupolas of large and small capacities with and without gases cleaning systems, with hot or cold blast.

In order to meet the different requirements of cupolas, different types of recovery system were considered.

The recovery plant consists as a rule of two subsystems:

- thermal subsystem including gas burner, combustion chamber, boilers with superheater and economizer, and steam drum;
- conversion subsystem for converting thermal into electric energy consisting of turbine-alternator, condenser, deaerator and circulation pumps.

The problems in planning the recovery system are connected with the cupola operating characteristics.

The melting plant is usually operated intermittently on an average of 16 h/day with some stoppages.

The gaseous emissions contain a great quantity of dust if a cleaning system is not provided, in our particular case there is a dust wet collector and the gas is cold, wet and moderately dusty. All these conditions made it necessary to go more deeply into the effects of quick start-up, load changes, fouling, dirt, corrosion of heat transfer surfaces and gas combustion parameters.

2. PILOT PLANT

2.0. Pilot plant concept

Before starting the construction of the industrial plant it was considered necessary to operate a reduced scale pilot plant in order to check the plan concept for the recovery system. The pilot plant concerns the thermal subsystem only, not the electric conversion one, because for the latter no special planning problems were foreseen.

The pilot plant principally includes a burner, a combustion chamber and waste heat boilers with superheater: a water droplets separator is also provided for use when a wet dust collector for the cupola gas emissions is provided.

First of all the pilot plant is designed to permit testing of the group consisting of the cupola gas burner and the eventual flame sustaining burner and affords a greatly approssimated determination of the combustion parameters and their effects on the provisions to be made for the complete definition of the industrial recovery plant design.
These parameters are for instance: excess air, combustion temperature, fuel (natural gas) consumption, requirement (or non-requirement) of preheating the combustion air or gas. All these parameters must be verified by taking into account the instantaneous actual contents of carbon monoxide in the cupola emissions.
In this way the waste heat boiler section can be tested in order to verify all the foreseeable problems concerning dirt, fouling and corrosion of smooth or extended heat transfer surfaces.

The tests were programmed in the real operating conditions of the cupola, but on a reduced scale (1 to 60 about). Since superheated steam can be produced, information will be obtained from testing about the thermal subsystem's ability to follow load changes in the start-up phase, during shut-down and following the daily operation diagram.

Without going into details, irrelevant to this research, considerable variations in carbon monoxide contents and consequrntly in gas heating power are possible. The dust in the fumes may have varying proportions of acid and basic oxides and thus a different refractoriness in each case with different effects concerning dust and fouling.

2.1. Design parameters and features

As described in the previous points the pilot plant includes the thermal subsystem only: the produced steam burning the CO gases is condensed in an aircooled condenser and the condensate is then recycled to the drum.

It has been designed according to the main following data:

		Gas Side	Steam Side
. Burnt gas flow rate	Nm3/h	400	
. CO contents	% Vol.	15 ÷ 22	
. Excess of air	%	10 ÷ 60	
. Thermal duty	Kcal/h	200 ÷ 300.000	
. Sat. steam flow rate	Kg/h		250 ÷ 300
. Superheated steam flow rate	Kg/h		200 ÷ 250
. Steam pressure	Bar		31
. Flue gas temperature			
. combustion chamber	°C	900 ÷ 1.100	
. to the stack	°C	280 ÷ 300	

The pilot plant main components are as follows:

. combustion chamber with CO burner and methane burner;
. thermal shield;
. steam superheater;
. vertical fire tube boiler;
. stack;
. steam drum;
. aircooled steam condenser
. circulation and recirculation pumps;
. CO and air fans;
. water droplets separator;
. water tanks;
. piping and instrumentation.

The CO gas is burned in the combustion chamber and the hot flue gas is circulated across the thermal shield and the steam superheater, there it is circulated into the tubes of the vertical fire tube boiler; after that the gases are discharged to the atmosphere through the stack.

The steam generated in the boilers leaves the steam drum, it is superheated in the superheater, then it is condensed in the aircooler and the condensate is recirculated again to the drum.

A pump provides to supply make-up water to the steam drum when the water operation level should fall below the minimum operation one because of leakages or intermittent blow-down.
Two pumps were also provided in order to help, if necessary, the water circulation through the boilers during transition periods like start-up, quick load variation etc.

Fig. 1 shows the flow sheet of the pilot plant.

2.2. Mounting

The pilot plant has been mounted at the Teksid Foundry of the FIAT Group, located in Crescentino far away 40 Km. from Turin.

The plant has been installed just outdoor of the industrial building where there are the 40 ton/h cast iron cupolas of the foundry.

The mounting has been performed in two stages as provided by the general test program: during the first step the combustion chamber with burner, gas and air fans and relevant instrumentation have been mounted. Several combustion tests have been then carried out.
After the combustion tests also the steam boilers and the plant balance have been mounted.

Each cupola produces different kinds of cast iron (grey and spheroidal): the pilot plant has been connected with both of them to perform a more extensive testing program.
To take CO gas, the plant was connected to each cupola just after the dust collector.

Piping and tube supports have been designed according to the Spande pipe-stress computer program.

After the mounting of the whole plant an air and soap test has been carried out on the steam-water side to minimize leaks.

2.3. Instrumentation and measuring equipment

The pilot plant has been instrumented in a semiautomatic way.

The CO gas and air combustion flows are manually regulated.
Should the CO gas be interrupted because of cupola shut-down, a quick-automatic valve isolates the CO gas line from the pilot plant; a flame detector actuates the same valve and closes the gas supply line in case of flame burn-out in the combustion chamber.
A semiautomatic sequency provides to start again the combustion.

The water-steam side operates in a semiautomatic way too; the operation steam pressure is manually selected by a switch, then it is self regulated at the set point value by a PIRC on the steam drum that controls the steam pressure by acting on the fan blades orientation of the air cooled condenser.

Two safety valves on the drum and one on the superheater outlet are provided, should the steam pressure incidentally overcome design level.

The relevant instrumentation (indications, alarms and emergency pushes) are situated in three adiacent panels:

- general panel for power connections of the plant, for pump, fans start-up and stop;
- control panel for CO and auxiliary CH_4 burners;
- panel for water-steam, water levels, pressure, set points, temperatures, etc. of the system

The instrumentation and valves can be considered of two different kinds of service: safety and operational, both for CO gas and steam sides.

Before starting the operational test of the plant, pre-operational tests of instruments and safety chaines have been carried out.

The pilot plant provides the following measuring equipment:

. CO gas flow meter, orifice type;
. air combustion flow meter, orifice type;
. methane flow meter;
. thermocouples for temperatures measurement in the combustion chamber (at the bottom, middle and top);
. thermocouples for temperatures measurement across thermal shield, superheater, fire tube boiler and at the stack;
. probes for measuring pressure drops across thermal shield, superheater and fire tube boiler on the gas side;
. thermocouples for measuring steam temperatures in the drum and superheater;
. steam pressure indication probe;
. condensate temperature probe;
. steam flow meter;
. steam drum and condensate tank level indicator and transmitter

3. TESTS PERFORMANCE AND RESULTS

As previously described the pilot plant has been mounted and tested in two steps.
At the first stage only the combustion group has been mounted and the combustion tests have been carried out; at the second stage the boilers and balance of the plant were also mounted.

3.0. Preliminary tests

Before operating with the pilot plant an extensive measurement program has been performed on the characteristics and analysis of the waste CO gas to be burned and entrained dust, in order to have a first check of the design parameters and economical provisions for pilot plant as well as for industrial plants.

3.1. Operational tests

As above mentioned the operational tests have been performed in two steps:

. First stage: during this step the combustion tests have been performed the plant has been in operation four weeks and two different kinds of burners have been tested.
The tests have been carried out during the 16 hours daily operation cycle of the cupolas.
During the night, when the cupolas are not in operation, the combustion chamber was hold hot in order to avoid thermal shocks on the refractory by a small supply of methane.

. Second stage: the plant mounting was completed with the part relevant to water-steam side including boilers, superheater, condenser, piping, etc.

The plant has been in operation burning the cupola waste gases five weeks.
During the night the water-steam side has been maintained in pressure by a small supply of methane.
Besides re-checking combustion parameters, during this step the boilers design, heat transfer characteristics, fouling, all the system behavior were tested.

3.2. Results

The main results obtained by the operational tests are of different kinds and summarized as follows:

Step one: combustion tests

- two burners of different designs were tested: the operation has been satisfactory for both of them;
- the wet and cold CO gases have been burnt with and without a rich fuel support;
- the influence of air combustion preheating has been checked;
- the range of combustion temperatures has been measured and compared with the theoretical ones;
- the gases and air flow rate were measured, besides the analyses of cupola gas CO contents and combustion products (i.e. O_2 contents) have allowed to check the air excess;
- the combustion chamber design has been verified with reference to the specific thermal load and refractory features;
- the effectiveness of droplets separator on CO gas before combustion has been tested and the amount of separated water measured

Step two: boilers and entire plant tests

- the boiler design has been checked;
- the fouling effect of the combustion gases has been checked and information about the components design was collected;
- the boilers thermal rate has been checked and compared with design calculations;
- corrosions on the CO gas line have been observed;
- the plant behavior has been verified during the operating cycle, start-up, shut-down and during quick load changes and sudden stoppages of cupola;
- the experimental tests have allowed to verify the variation influence of the metallic charge, coke and O_2 percentage added to the hot blast, of the cupola on the CO contents of the waste gases

Fig. 2 shows the combustion chamber temperatures vs. CO contents in the waste gas.

Fig. 3 shows the combustion chamber temperatures vs. % and type of coke in the cupola charge.

In Table III the powder analysis found on the boiler tubes is shown.

4. ANALYSIS OF RESULTS

We outline some relevant results of the tests performed.

Combustion tests

This class of tests have been very important because they have allowed to check the following points:

- the wet CO gas at 30°C can be burnt in the refractory combustion chamber; the chamber optimum specific rate of combustion has been evaluated;
- no methane support flame by natural gas is needed;
- no air preheating is required to burn the wet CO gas;
- the minimum excess air required has been measured and it is higher than previously provided at the design stage;
- the combustion temperatures have been measured and compared with calculations;

The combustion tests have given basic and very important information: they have confirmed the theoretical absumptions, but also given a more exact indication about the design parameters.

The minimum excess air found to be necessary during the tests has been about 40% instead of 10% as per the previous design and in certain operation conditions they can reach 50%.

Boiler system behavior

The boiler concept has operated very satisfactorily in forced boiling water circulation and also in natural water circulation.

The thermal rating and the pressure drops have been checked and compared with theoretical approach in order to find the most suitable rating for indutrial plants.

The fouling on the heat transfer surfaces has allowed to operate the system for two weeks after that nearly the calculated conditions were observed.

However for the industrial plant the accessibility and the in-service cleaning capability of the boiler sections must be improved.

Fouling

After two weeks of full operation with cupola gases at design conditions, some fouling was observed.

The fouling consists of a powder continuous layer on the surfaces about 0.1, 0.2 mm. thick. that adheres very hardly to the heat transfer surfaces.

The powder is composed of very small grains of brown-white colour.

The fouling has been higher than provided at the design step for the described operating conditions and it is also more difficult to remove: for that reason the accessibility to the surfaces to be cleaned on the industrial plant is to be improved and more adequate cleaning devices are to be provided.

5. CONCLUSIONS

The scope of the test programme stated at the beginning was to demonstrate the technical and economical feasibility of the above described kind of recuperation systems burning cupola CO wet gases.

For many reasons among which mainly required for economy and time sake it has been decided to test the new concept and new application of the thermal subsystem only because the electrical conversion one is more conventional.

The tests have been carried out in three subsequent periods as described (combustion, boiler performance and fouling, superheater and operating cycle) for a total operational time of about three months.

All the components of the thermal subsystem have been tested (burners, combustion chamber, boilers, superheater, gas water droplets separator, instrumentation), as well as the general concept and layout.

As a conclusion the experimental tests confirmed also the technical and economical feasibility of a recuperation system using a low energetic gas as the one flowing out from the cupola, the same being collected after the wet gas dedusting system.

DISCUSSION

Question - Dr. Thornton, Courtaulds : One of the strong points of your process is that maximum power output of the unit corresponds peak power demand of the foundry.
This is a keypoint, and problems might occur if the correspondence were not exact. Are you confident that the system will work in practice ?

Answer

1. The recovery system burns the gases at the exit of the cupola that melts the cast-iron.
The cast-iron lines as well as the drafts for blast, gas cleaning etc. operate for a long time simultaneously with the electric power production. However we have assumed conservatively that the operation/demand ratio will coincide for 95% of the time.

2. As to your second question, we have observed that the pilot-plant (on the scale 1:60 of the industrial plant) has operated without any problems for about 3 months.
About the scaling effect, we can observe that the combustion on a larger scale should be better than on the pilot-plant as the capacity of the pilot-plant corresponds to the minimum size of the burner that could be designed for that service.
The boiler behaviour should be also better as we have improved some design parameters according to the test results.

TABLE I

CUPOLA OPERATING DATA

Melting rate	a) maximum	40.000 Kg/h
	b) present	26.000 ÷ 36.000 Kg/h
Blast	a) flow rate	19.000 ÷ 25.000 Nm3/h
	b) temperature	500 ÷ 510°C
Gas temperature	a) charging door	235 ÷ 280°C
	b) before cleaner	220 ÷ 265°C
	c) after cleaner	40 ÷ 50°C
CO contents	a) charging door	16 ÷ 22% vol
H_2O contents	a) charging door	0 ÷ 2%
	b) after cleaner	9%

TABLE II

RECOVERY SYSTEM MAIN FEATURES

		Cupola furnace of 40 t/h (°)
Emitted gas flow rate	Nm3/h	25.000
Gas temperature	°C	40 ÷ 50
Average CO contents	% vol.	19
Total heat recovered	Kcal/h	12×10^6
Electric power output	Kw	3.500
Electric energy produced/year	Kwh	12.3×10^6
Obtained Kwh/ton of cast iron produced		87.5

(°) Cupola furnace with effluent gases wet cleaning system

TABLE III

POWDER ANALYSIS FOUND ON THE BOILER TUBES

Main components:

S_1O_2	23 %
Fe_2O_3	29,8 % (expressed as iron)
Al_2O_3	4,6 %
CaO	4,2 %
MgO	1,07%
Na_2O	1,15%
K_2O	2,5 %
H_2O	5,8 %
CO_2	0,2 %
C	1,84%
S	2,37%
MnO	2,65%
CnO	0,14%
ZnO	5,7 %
PbO	0,55%
SnO_2	0,44%

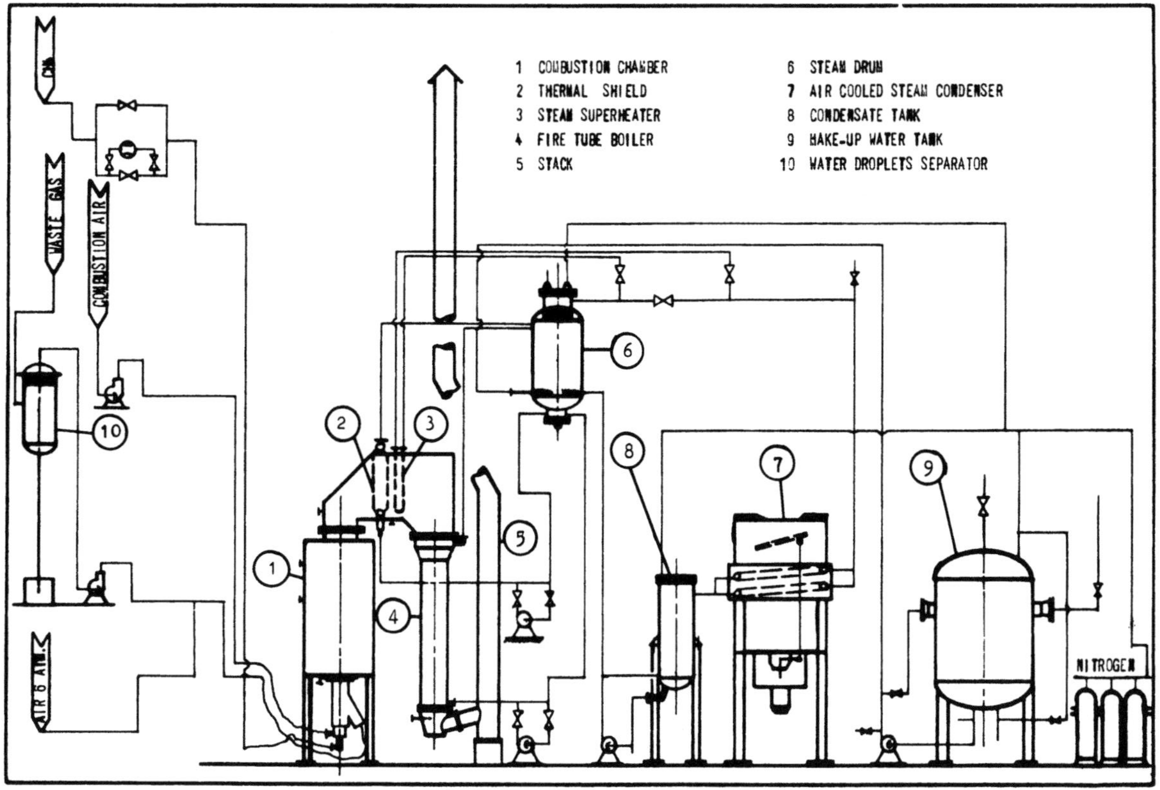

Fig. 1 Pilot Plant Flow Sheet

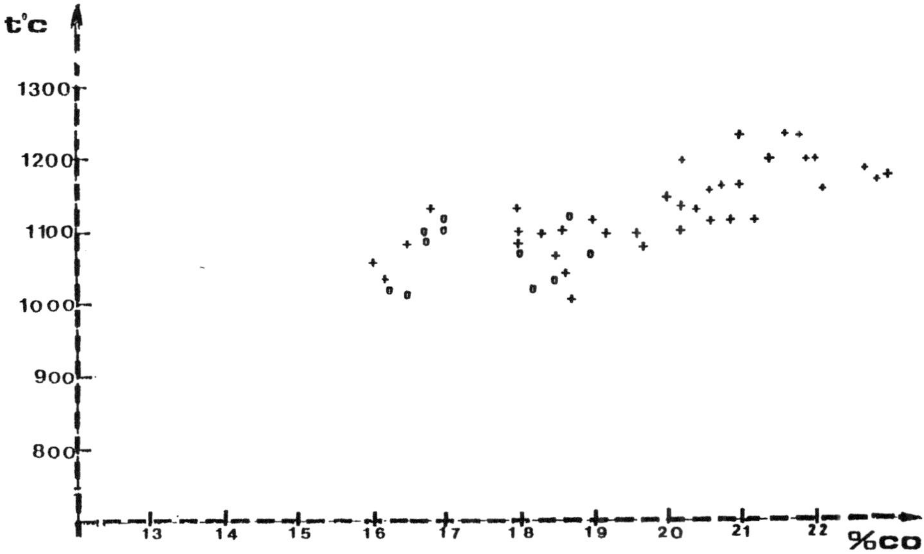

Fig. 2 Combustion chamber temperatures v.s. CO content in the waste gas (radiation effects included)

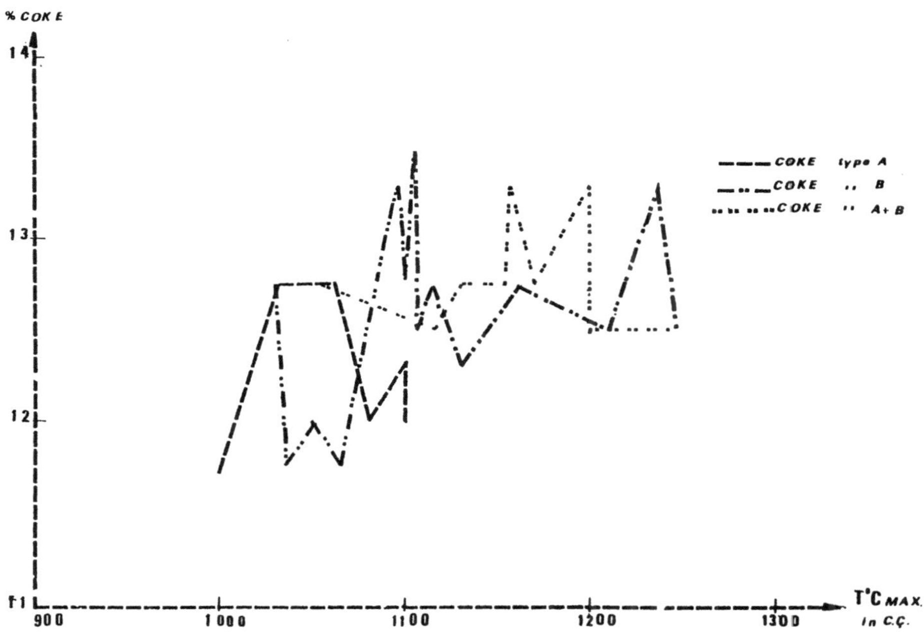

Fig. 3 Combustion chamber temperatures v.s. % and type coke in the cupola charge

SESSION III - ENERGY CASCADING AND ORC ENGINES

Chairman: J. LIMIDO

SUMMARY OF SESSION III

ENERGY CASCADING AND ORC*) ENGINES

J. Limido**), P.A. Pilavachi

The research work carried out under the above title has as an objective to investigate the rational use of thermal energy by improving the overall thermodynamic efficiency between the temperatures of the generated and the final rejected heat. A wide range of industrial processes could thus benefit from the outcome of this work.

Within this framework, four projects concentrate on the use of a Rankine cycle. These are applied either as a bottoming cycle (3 projects) or as a topping cycle (1 project). Furthermore, one project assesses the difficult problems encountered in industrial energy management which combine cascading thermal energy and its storage, and concentrates on the case study of the United Kingdom.

Industrial waste heat recovery - The project undertaken by M.B.B. is a combined analytical and experimental programme to investigate the feasibility of applying the ORC principle to waste heat recovery in industry. The temperature ranges of 200° C to 400° C will be given special attention in the investigation since for these upper temperature ranges only a limited amount of information is available on the behaviour of the fluids and the adapted technical components in the ORC. The economy of waste heat recovery in an ORC depends greatly on the selection of the appropriate working fluid because its cost factor could be nearly half of the investment costs for the total plant.
Using test facilities which exist at M.B.B. realistic dynamic tests can be achieved, to complete the stability reaction laboratory tests and to compare the different types of components (mainly expanders).

Medium temperature heat recovery - The aim of the research work done by TURBODEN s.r.l. is to assess the possibility of building at a reasonable cost a 100 kW ORC unit. The application of this unit would be valid both for industrial waste heat recovery or for cogeneration plants, which discharge heat at temperatures well above those usually required by heat users (as is the case with Diesel engines or gas turbines). The performance tests will be carried out at the TURBODEN specific test plant facilities near Milan. The design and specification of the different items of the engine are finished and construction is almost complete.

*) ORC = Organic Rankine Cycle
**) IFP-ENSPM, France

Multivane expander - The work conducted by the CRANFIELD INSTITUTE OF TECHNOLOGY is directed at developing low-cost, robust, quiet and efficient multivane expanders (MVE) as prime movers in ORC. It was concluded that MVE performance is primarily affected by its design with respect to mechanical friction and leakage. Furthermore, an ORC may be optimised for maximum efficiency once the working fluid and expander characteristics are known; if an effective positive internal sealing is employed, high efficiencies can be obtained.

Sulphur topping cycle - The overall efficiency of a binary sulphur and water cycle plant is estimated at some 60 % according to studies carried out by the COMMISSARIAT A L'ENERGIE ATOMIQUE (C.E.A.). The present contract covers the experimental confirmation of the technical feasibility of the topping sulphur section. The problem of corrosion of steel by sulphur at high temperatures had been solved at a previous stage. Now the C.E.A. is in a position to establish the required data for the properties of sulphur at high temperatures and to test the behaviour of the materials under real plant conditions. There are technological difficulties in the construction of the loop and preliminary tests are necessary. The main components of the loop have been ordered and the detailed study of the design is in progress.

Energy cascading and industrial heat storage - The potential to apply energy storage cascades within the UK industry was assessed by the CRANFIELD INSTITUTE OF TECHNOLOGY. The work includes both the technical and economic aspects. The object of the study is to estimate realistically the industrial potential of storage cascades. The identified regions of application of this type of storage system reside within industrial sectors where conventional waste heat recovery would, in many cases, be an economically superior option to storage. In addition the greatest potential for steam storage does not lie in associating it with an industrial process, but in integrating it into the steam service system.

INDUSTRIAL WASTE HEAT RECOVERY BY USE OF ORGANIC RANKINE CYCLES (ORC)

Author : G. Huppmann

Contract Number: EE - B - 1 - 108 - D

Duration : 30 months 1 December 1981 - 30 May 1984

Head of Project: Dipl.-Phys. G. Huppmann, Messerschmitt-Bölkow-Blohm GmbH
- Space Division -

Contractor : Messerschmitt-Bölkow-Blohm GmbH
- Space Division -

Address : Messerschmitt-Bölkow-Blohm GmbH
P.O. Box 80 11 69
D-8000 München 80

Summary

The project is a combined analytical and experimental programme to investigate the feasibility of the Organic Rankine Cycle principle for waste heat recovery in industry. In this report the main objectives and the work programme are described.
Some analytical investigations have preliminary results, such as our study about the potential of waste heat in industry usable by ORCS and the determination of the three most typical ORC-plants to be planned more detailed. There will be one plant in the petrochemical industry, one in the ceramic industry probably and one for using the thermal energy of Diesel-engine exhaust gases.
First results of the investigations in the experimental programme are given, too. Some details about thermal stability of R 114 can be described as well as the experience gained with the test loop during the experiments with R 114.
The characteristics of the different heat exchangers in the test loop are discussed and compared with theoretical values.

1. The main Objectives of the Project

Our project is a combined analytical and experimental programme to investigate the feasibility of the Organic-Rankine-Cycle principle for waste heat recovery in industry.
There is a lack of information about the potential of industrial waste heat usable by ORCS as well as the application of organic fluids as a working media in an ORC. Since there is only a limited amount of information available in the relevant upper temperature ranges concerning the behaviour of the fluids and the adapted technical components in the ORCS, we concentrate our investigations in these ranges.
The economy of waste heat recovery in ORCS depends greatly on the selection of the appropriate working fluid because its cost factor could be near the half-mark of the investment costs for the total plant.
Therefore we have to investigate and compose the criteria for determining the most suitable fluid for any application in an ORCS. On this basis, the representative items of design of the technical construction of the whole ORCS and its components are given. An ORCS and component analysis is conditional for the definition of the data, which can thereafter be used as a basis for an extrapolation to other power ranges interesting.
A further essential point in our investigations is the economical description of the waste-heat-recovering-ORCS, together with a cost sensibility study.

2. Work Programme

2.1 Analytical Investigation Programme

Our analytical investigation programme consists of the following activities:

- The identification of typical operational requirements for the working fluids, for the representative items of design and the cycle and components analysis
- the collection and evaluation of basic data for the design of ORCS for any waste heat recovery in industry, and the execution of the planning of the three most typical ORCS plants
- the collection of data concerning the potential of industrial waste heat usable by ORCS, and the specific part of the energy used by ORCS as compared with the total of the energy consumption in the industry.
- the investigation of the economical constraints in order to evaluate the economics and the cost sensitivity of design parameters.

Some results of these activities can be given, as follows:

Requirements for the fluids and lubricants

- Thermophysical Requirements

One of the fundamental criteria for the design of an ORCS is the maximum waste heat temperature. The selection of the working fluid depends mainly on this criterion. In order to have a low system operating pressure the working fluid should have a relatively high boiling point in comparison with the maximum waste heat temperature.
Another desired property is a good heat transport coefficient in order to minimize heat-exchanger surface areas.
A third criterion is the slope of the saturated vapour line. Organic fluids with a high molecular weight and a low number of atoms per molecule have positively sloped saturated vapour lines. Therefore, the vapour of these fluids is superheated at the exit of a turbine, and requires a recuperator to increase the system efficiency.

The vapour of the fluids with negatively sloped saturated vapour lines (e.g. water) is emitted - usually in a moist state - by the turbine. The ideal fluid for a Rankine cycle power system has near-vertical saturated vapour lines, and this is the reason for mixing negatively sloped fluids with positively sloped fluids, e.g. Trifluoroethanol with water (Fluorinol) (1)

- Thermal stability

The selected fluid should be thermally stable at all operating temperatures. Especially the mixing of the fluid with a usable lubricant has often presented problems with the thermal degradation.
There is a lack of information on these points with regard to higher temperature ranges, and we hope that we can give you this information in the near future for R 114, Toluene and for Fluorinol 85.

- Material compatibility

The selected working fluid must be non-corrosive when in contact with the common engineering materials used in an ORCS at all operating temperatures.

- Health and safety requirements

In order to avoid accidents the selected working fluids should be non-toxic, non-inflammable and non-explosive otherwise respective safety installations have to be introduced, such as explosion safety devices or gas sensors and alarm systems etc.

Potential of waste heat in industry usable by ORCS

A study is initiated about the potentials of the waste heat in industry, which can be recovered by use of ORCS. The following data are available to date:

- Petrochemical industry

Fig. 1 gives a general view on the potential application cases of ORCS in German refineries.

ORCS performance kW_{el} \ process	atmosph. distillation	vacuum distillation	hydro-finer	thermo cracker	catalyt. cracker	plate-former	total
100 - 500	-	22	20	4	15	42	103
500 - 1000	5	16	49	6	6	14	96
1000 - 1500	12	11	11	2	2	6	44
1500 - 2500	26	5	1	2	1	7	42
2500 - 4000	11	-	-	2	1	-	14
total	54	54	81	16	25	69	299

Fig. 1 Potential for ORCS in refineries at different power ranges (Federal Republic of Germany)

It can be seen that the waste heat recovery of a hydrofining plant is a typical ORC-plant in the petrochemical industry in the medium power range, followed by plants for atmospheric distillation.
The overall potential of waste heat for use in ORCS in the petrochemical industry seems to be very high, however, we do not expect excellent economic basic data because of the high level in the energetic integration of refineries. The waste heat is used frequently for preheating feed material, or as process-energy.

- Iron and steel industry

The most important waste heat source in the iron and steel industry is the sintering cooler.
Sintering of ore is a necessary step to avoid the clogging of the blast furnace. In the sintering process the ore is heated between 550°C to

650°C, and then it is cooled to storage temperature. The sintered material is cooled by fresh air, which streams through the hot porous material.
The air can be collected by capsulating the cooler and is used as a heating medium in the vaporizer of an ORCS.
Roughly speaking, there are today approximately 50 plants in Europe, which are suitable for ORC-waste-heat-recovery.
The total electric energy output would be approximately 350 MW, and the typical performance of the single plant would be in the range of 2 MW to 10 MW. The investment costs for installation of waste-heat-recovering ORCS in sintering plants differ as the installations for collecting hot air stream, and its dust filter depend on the individual design of the sintering plant.
One of the typical plants is the sinter-cooler of Thyssen AG in Schwelgern. It has a cooling area of approx. 400 m², a capacity of 4,6 Mio tons, and ranges within the 25 biggest plants in the world. This plant would permit the installation of an ORCS with an electric output of approx. 11MW.

- Diesel Engines

One of the most interesting potential of waste heat usable by ORCS are the exhaust gases of diesel engines. The maximum temperature of the exhaust gases behind the valve is about 540°C. If you accept a certain temperature loss on the way to the vaporizer of the ORCS you can use Fluorinol 85 or Toluene as its working fluid. The additional performance of the ORCS is taxed on 25 % of the original mechanical performance of the engine. The total mechanical efficiency increases from 35 % to 43 % by using an ORCS.
Fig. 2 shows the potential for the use of ORCS in Diesel engine-driven ships.

Engine power range (kW)	produced in 1980	mean power increase by ORCS (kW)
100 - 300	587	50
300 - 500	180	100
500 - 1000	229	190
1000 - 2500	668	400

Fig. 2 Potential for ORCS at main Diesel engines of ships in Europe

The potential of ORCS on stationary Diesel engines is not very high, in our opinion, as the waste heat of such engines can be used as thermal energy for heating.

- Other industrial potentials

We are collecting data about other waste heat potentials usable for ORCS. They are to be found:
- at gas turbines
- in the ceramic industry
- in the cement industry
- and at some industrial furnaces.

Selection of typical ORC-plants

Information of some typical waste-heat-recovering ORC-plants can be derivated from the presented data collection.
Considering different waste heat media and temperature ranges in comparison with different working fluids we selected the typical plants, as shown in Fig. 3.

waste heat producer	waste heat medium	possible working fluid	ORC-performance expected (kW_{el})	potential application number
Petrochemical Industry - waste heat from hydro-fining	condensing mixtures of hydrocarbons ca 230 °C, ca 3.75 MW	Fluorinol 85 or toluene	ca 300	49 in the power range of 300 - 700 kW (Germany)
- atmosperic distillation	fluids, 160 °C	R114	1500 - 2500	26 in Germany
Diesel engine of ships	exhaust gases with SO_2 and carbon black max 540 °C engine performance 1000 - 2500 kW	Fluorinol 85 or toluene	200 - 600	600 p.a. in Europe
Iron and steel industry	air with dust from sintering-plants 340 °C	Fluorinol 85 or toluene	14 MW	50 in Europe

Fig. 3 Basic data of typical ORC-plants

We have done more detailed planning on the example of the waste heat recovering ORC for the hydrofiner.
This ORC is based on the data of the hydrofiner of the ESSO Corp. in Ingolstadt. It uses the waste heat of the tailgas condensor and the treatgas cooler. The waste heat input is approx. 2.6 MW (see Fig. 4), the electric power output approx. 260 kW.

The investment costs are taxed at approx. 1 Mio DM, and the return of capital will be 4 years, considering 10 % interest, 6 % p.a. increase of the energy price of 0,16 DM/kWh.

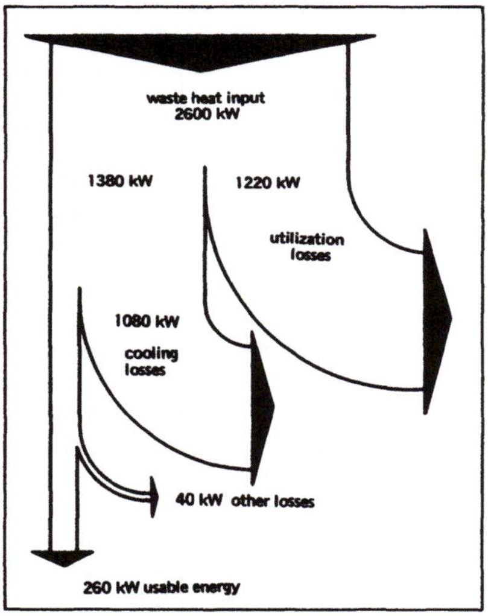

Fig. 4 Flux of energy
Waste heat recovery from a hydrofiner by ORCS

2.2 The Experimental Programme

The main tasks of the experimental programme are:
- the thermal stability of usable organic working fluids in the necessary mixture with adapted lubricants up to the temperature range where water is used as a working fluid.
- the experimental evaluations of thermodynamic data of the organic work-

ing fluid, such as
- o heat transfer capacity
- o evaporation energy
- evaluation of cycle and component performance compared with theoretical design values.

We concentrate this test programme on three typical fluids for ORCS
- up to approx. 200°C, the refrigerant R 114
- up to approx. 300°C, the Fluorinol 85, a mixture of Trifluorethanol with 15 % water
- up to nearly 400°C, the Toluene.

The experiments with each fluid consist of two different test stages. For evaluation of the loop and component performance and of the thermal stability of the fluids we test them in our experimental Rankine Cycle under realistic operational conditions.

In this test loop, we increase the temperature of the superheater following a given timetable, and take samples of the fluid/lubricant mixture to check the rate of degradation products. These samples were analysed concerning these defined degradation products which provide the best information about the further usability. In the case of R 114 we analysed the contamination by the chloride ion and the fluor ion.

Other parameters are the conductivity and the concentration of hydrogen (pH-value).

The other experiments are the reaction kinetic tests under static conditions. We store the fluid/lubricant mixture in test containers at different temperatures during different periods to evaluate the degradation rate of the mixtures in contact with the container's material. From this degradation graph we try to find the correlation to the graph of degradation of the dynamic experiments. May be, this correlation makes it possible to provide extrapolated information with a fair probability about thermal stability of any fluid/lubricant mixtures in a realistic ORCS on the basis of static tests only.

The degradation rate is identical with the reaction velocity and means the decrease of modes per time unit. In general, the reaction velocity is a function of the concentration (c) of the reactants and the temperature (T). It can be equated, as follows:

$$-\frac{dc_{fluid}}{dt} = F(c_{fluid}, c_{oil}, c_{Fe}, c_{FeO_x}, c_{Me}, T)$$

(with c_{Me} = concentration of catalytical aging metals)

This equation can also be written as:

$$-\frac{dc_{fluid}}{dt} = K(T) \cdot f_1(c_{fluid}) \cdot f_2(c_{oil}) \cdot f_3(c_{.....}) \text{........}$$

K (T) is a function of temperature, and is called the velocity constant. By the equation of Arrhenius

$$\ln K = C \cdot \left(-\frac{1}{T}\right) \quad \text{or} \quad K = K_0 \cdot e^{-\frac{E_A}{RT}}$$

the factor K_0 is independent of temperature, and E_A is the activation energy for the reaction.

If K_0 and E_A are known, it is possible to calculate the velocity constant for a fixed temperature, and after this, the reaction velocity

$$-\frac{dc.}{dt}$$

It appears possible to define a correlation function between K_o of the static experiments to the K_o of the dynamic experiments under duty cycle conditions, and we hope that we can give predictions about the thermal stability of working fluids on the basis of static storage experiments only.

2.3 Description of the Test Facility

MBB built some ORC-plants in the last years. We started with a 20 kW pilot-plant for the low temperature range. A screw expander is installed in this plant. The plant is designed for using the thermal solar energy from flat plate collectors for water pumping and other similar purposes (Fig. 5).

Fig. 5 20 kW pilot ORC-plant

Another plant is built for using the thermal solar energy collected from a parabolic dish field. This ORC works with Toluene and is designed for maximum temperatures of approx. 400°C. The prime mover in this plant is a single stage radial turbine. The electrical output at the design point is about 150 kW (Fig. 6).

Fig. 6 150 kW ORC-plant

The experience with these plants were our starting point to initiate the mentioned programme for investigation of organic fluids for Rankine power systems.

For testing organic fluids under closed loop Rankine Cycle conditions a test facility has been designed and put in operation at MBB's test area in Ottobrunn in 1979.
The Rankine loop of this test facility consists of the following main components:

- vapour generator (preheating, evaporation and superheating)
- prime mover with lubrication loop (2 types of screw expanders and a transsonic turbine)
- throttle valve
- recuperator, condenser and subcooler
- feed pump

The main characteristic data of test stand are the following:

- thermal heat input	200 kW
- maximum temperature	400 °C
- heat transfer medium (heat source to vapor generator)	Diphenyl/Diphenyl Oxide
- max. mechanical power output	25 kW_{mech}
- variable drive	2000 ./. 10000 min^{-1}/1500 min^{-1}
- alternator	50 Hz/32 kVA, 1500 min^{-1}
- waste heat removal	water cooling

The figures 7, 8 and 9 show some details of the plant.

Fig. 7 The ORC testing plant. The upper cylinder is the vaporizer, below that the recuperator and the tank for the working fluid. On the left the fluid/lubricant separator and on the ground the feedpump

Fig. 8 The screw-expansion engine installed during the tests with R114

Fig. 9 The 200 kW/400°C heater for simulating waste heat

The data acquisition system handles the data of 40 pick-ups.

22 temperatures
8 pressures
6 flow rates
2 rotation speeds
1 torque
1 power

are evaluated.
The results can be presented tabulated or in diagrams (whatever is preferred).
By means of this data acquisition system it is possible to observe the test facility itself during operation and to measure the interesting data of each cycle component.

2.4 First Results on the Thermal Stability of R 114

The reaction-kinetic tests of R 114 under static conditions are still being evaluated, but we have finished the dynamic tests for a maximum temperature of 250°C because the installed screw-expansion-engine did not allow higher temperatures. The engine is designed for a maximum temperature of approx. 160°C. The overall testing time was 143 hours; above 220°C 23 hours.

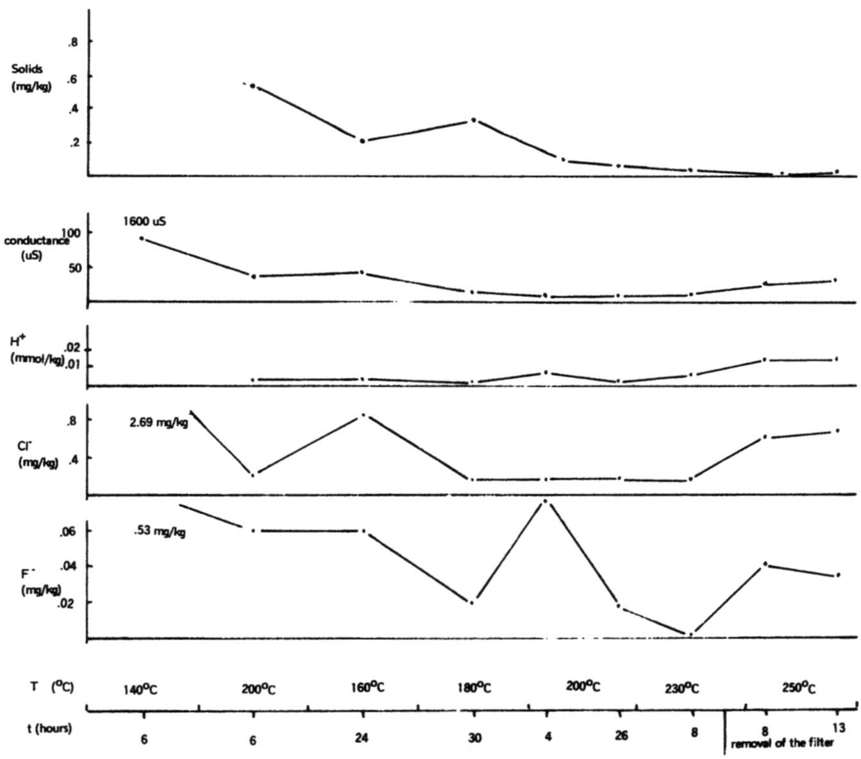

Fig. 10 Degradation of R114 / Clavus G 100

Below 230°C we could analyse only marginal decomposition rates, which were near the detection efficiency or the chemical analysing methods.
The marginal degradation rate is a consequence of the integrated filter system. However, this filter is a component of the claim to investigate under realistic conditions. The filter system removes the moist, solid components and acids.
On removal of the filter at an operating temperature of 230°C, the degradation rate increases, but the concentration of the decomposition products were still only traces.
It appears that the temperature limit of 160°C as described in literature (2) for the application of R 114 in ORCS, can be raised to 200°C.
Fig. 10 shows the cumulative concentrations of the degradation products versus the operational temperature and the testing periods.
It can be seen that the degradation products decrease at the beginning of the tests. This is a consequence of the filter, which removed many products generated during the long period of non-activity of the plant (nearly 3 years). After the 230°C point, the concentrations increase.
The measured values for H^+ and F^- at 195°C cannot be explained at this stage. It could stand for a high temporary concentration of HF, which decreases again at higher temperature ranges.

2.5 General Comments and Experience Gained with the ORC-Cycle

The primary purpose of the ORC-test-cycle at MBB it was originally designed for is to test a wide variety of working fluids and also single components as heat exchangers, expansion engines and so on. This has the consequence that this test cycle is far away from being an optimum design for a special fluid.

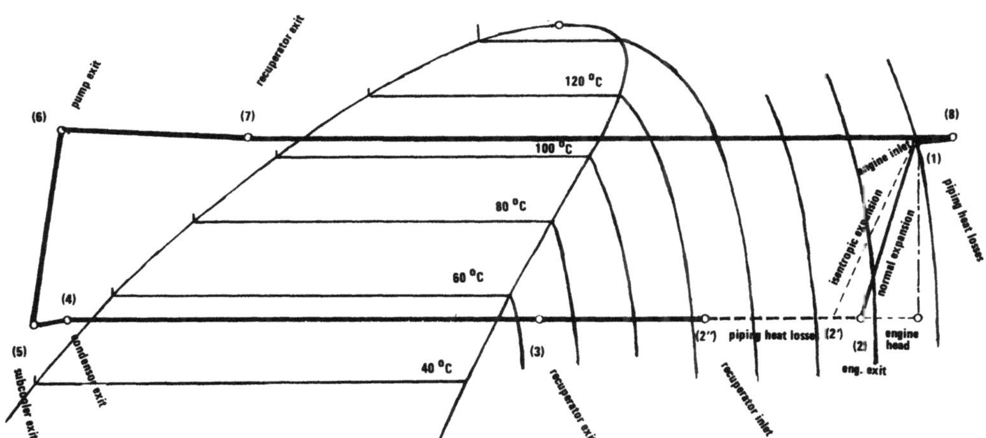

Fig. 11 Typical diagram of an experimental cycle for a particular data point

Therefore the cycle primarily can be used for the evaluation of performance and operational behaviour of the various components but the overall efficiency in terms of mechanical power output related to the thermal energy input is not at all representative neither for the fluid tested nor for the mechanical arrangement of the cycle.

Furtheron due to claimed accessability of components and sensors the effort on thermal insulation had to be limited below a level which would be desired for a commercially operating Organic Rankine Cycle. The operation of the cycle with R 114 showed some problems with the condensor which originally was designed for condensing working fluids at a temperature level of more than 100°C which is due to the fact that only a single stage expansion engine is used in the cycle. This again is due to cost reasons. A commercially operating cycle would require a two-stage-engine. For the operation with R 114 the nominal condensing temperature is only about 30°C. The result was that due to the limited capacity of the condensor the condensation temperature increased to about 60°C.

However, the primary test objective to evaluate the temperature limitations of R 114/lubricant was not effected by this problem. In the first test series all thermocouples were located as close as possible to the heat exchangers in order to give precise information about their behaviour. This had the consequence that the efficiency of the engine could not be measured as indicated in the cycle diagram of Fig. 11. The reason for this were the thermal losses of the pipes between superheater and expansion engine inlet as well as between the engine outlet and the recuperator inlet.
In the following test series where a direct comparison between screw expander and turbine is planned the temperatures will be measured immediately upstream and downstream of the engine.
With the exception of the problems just mentioned the test objectives of the test series with R 114 were fully achieved: the temperature limit for the operation of Frigene 114 in an Organic Rankine Cycle could be evaluated experimentally and can be compared to laboratory tests of the fluid. Furtheron the heat transfer behaviour of the heat exchangers could be examined and compared to theoretical values.

2.6 Test Evaluation of Vaporizer Unit

The test data for the vaporizer showed to be a very strong function of the heater exit temperature. It was not possible to obtain a single curve for all the data points. The data were therefore plotted for various vaporizer temperatures, as shown on Fig. 12. This diagram is very useful for extrapolation purposes, or for estimating the size of other vaporizer units.

Fig. 13 shows the experimentally determined overall heat transfer coefficient as a function of the flow rate. It can be seen in this case that there is considerable scatter in the experimental points, but the general trend can be seen. The theoretical overall coefficient was calculated and showed, as by the condenser, to be a strong function of the average vaporizer temperature. It was therefore necessary to display the theoretical curve as a band which encompasses all temperatures between 175°C - 120°C (i.e. average temperatures for the test data). The theoretical values were calculated according to the methods described in the "VDI-Wärmeatlas" heat transfer handbook.

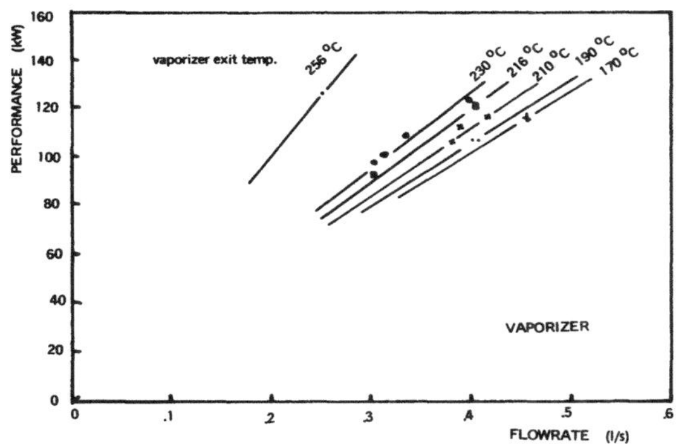

Fig. 12 **Vaporizer performance as a function of flowrate and various exit temperatures**

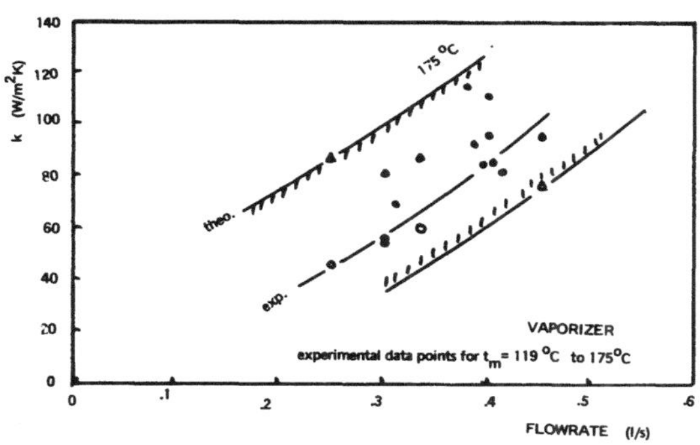

Fig. 13 **Experimental and theoretical over-all heat transfer coefficient as a function of flowrate**

2.7 Test Evaluation of Condenser

As previously mentioned, the condenser was much too small, when R 114 was used in the test cycle. This is clearly shown by Fig.14 , where the condenser performance is plotted as a function of the refrigerant flow rate. The design flow rate is roughly 0.4 l/s, which gives a condenser performance of 70 kW. Under similar conditions, a condenser performance of at least 80 kW would have been necessary to attain a desired condenser exit temperature of about 40°C.

Fig. 15 shows the experimental overall heat transfer coefficient K, as a function of the refrigerant flowrate. Although, some scattering of the data is apparent, a definite trend can be seen. This curve is very helpful for extrapolation purposes, or for estimating future needs. The theoretical overall coefficient was also calculated and showed to be a strong function of the average temperature of the condenser. It was therefore necessary to show a band, in which the theoretical values lie.

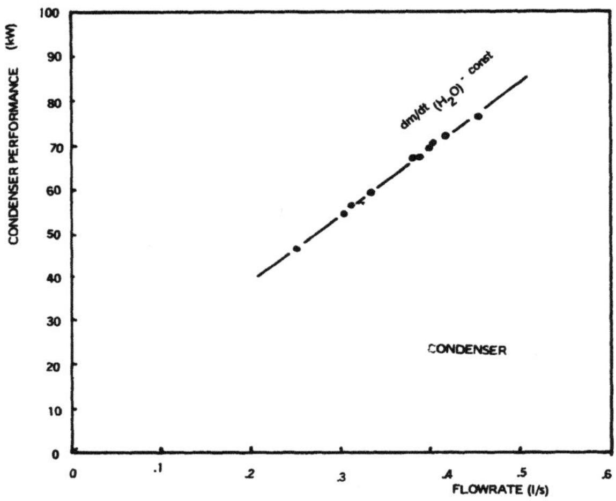

Fig. 14
Condenser performance as a function of flowrate

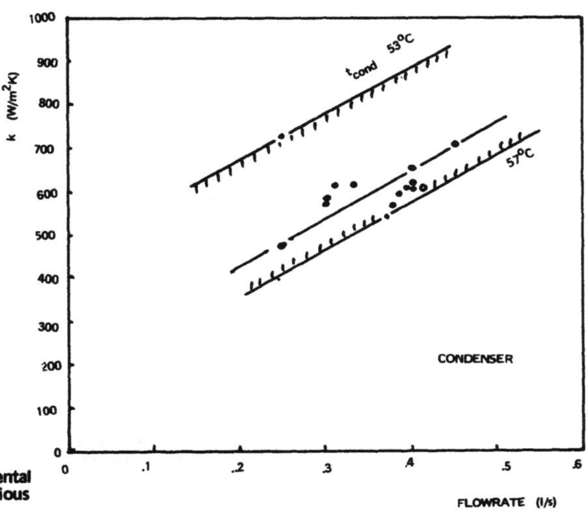

Fig. 15
Comparison of theoretical- and experimental over-all heat transfer coefficients for various flowrates

2.8 Test Evaluation of Recuperator

The recuperator is designated with points (2") - (3) (gas-side) and with (6) - (7) (liquid-side) in Fig. 11. As was done for the condenser, the recuperator performance was plotted as a function of the fluid flowrate in Fig. 16.

Fig. 17 shows again the experimental overall heat transfer coefficient of the unit, as a function of the fluid flowrate. The range of values can be seen to be one order of magnitude lower than the condenser values. The theoretical coefficient was also calculated, by the evaluating the of gas - and the liquid side of the recuperator. It can be seen that there is relatively good agreement between the two curves.

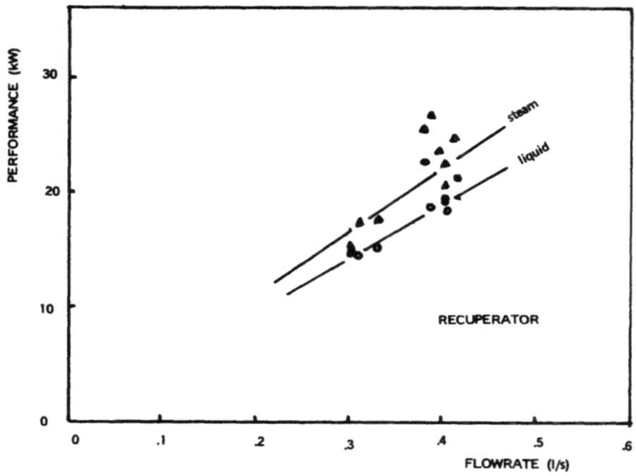

Fig. 16 Recuperator performance as a function of the fluid flowrate (gas- and liquid-side)

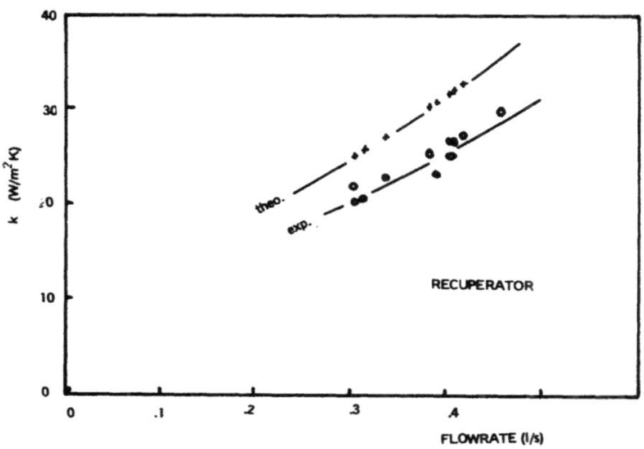

Fig. 17 Comparison between theoretical and experimental over-all heat transfer coefficients as a function of fluid flowrate

Literature

(1) Comparison of Rankine Cycle Power Systems: Effects of Seven Working Fluids
T.J. Marciniak et.al. NTIS, 1981

(2) Working Fluids
W. Scholten, VDI-Bericht 377, 1980

(3) Organische Arbeitsmedien in Wärmekraftmaschinen für Solarkraftwerke
Schmidt et.al. BMFT-FB-T 81-154, 1981

(4) Wärmerückgewinnung in Betriebsanlagen der Industrie
VDI Bericht 435, 1982

(5) Energetische Berechnungen von Kreisprozessen
Bitterlich et.al. BWK 34, 1982

(6) ORC zur Umwandlung von Abwärme in mechanische Energie
H.P. Corneille, Linde Bericht 46/1979

(7) Physikalische Daten für Frigen 114
Hoechst AG, 1980

DISCUSSION

Question - Mr Gaia, Turboden/CNPM : Did you have any decomposition in the primary loop at 400° C ?

Answer No, because the highest temperature till now was about 300° C. In the next weeks, however, we will run up to about 395° C. Diphyl without lubricants is suitable for up to 400° C.

Question - Mr Gaia, CNPM/Turboden : What kind of filter did you use in your test loop ? Were you concerned with the possibility of introducing chemicals in the loop which would influence the thermal decomposition ?

Answer The filter was of the active carbon type. These filter systems will be used in any realistic ORCS because of prolongation of the usability-time of the fluid. The decomposition rate is held under a certain threshold by these filters which are placed in the cooler part of the loop, were the degradation products have a smaller reaction velocity constant.

Question - Mr Bell, Cranfield Institute : 1. Have you considered using a contaminant filter in an ORC loop to eliminate acids, water, corrosive products etc. ?
2. Please measure the isentropic efficiency of your expanders !

Answer 1. We used in the tests with R 114 a filter of active carbon to eliminate these products. In any realistic ORCS a filter like this will be used because it can prolongate the usability-time of the working fluid.

2. We will do it in the next test phase, where we test toluene.

Comment - Mr Wood, Cranfield Institute : With regards to ORCS used in chemical plants it is important to integrate ORCS into a lower temperature 'system' i.e. that which is a net heat source to the cooling utility. Because of the nature of many plants relatively low temperatures are available i.e. 120° C in the lower system. Care should be taken not to overestimate the potential in this area.

Answer We will consider it.

Comment - Mr Macchi, Politecnico di Milano/CNPM : I think that the fluid selection is mostly a matter of the power output of the engine. The fluid should have the right pressure of condensation to ensure a suitable flow rate in the expander. So R 113 should be used in low power output ranges, and R 114 for higher power outputs. We selected R 114 as working fluid in a 500 kW unit for geothermal hot water for this reason, while we used very low pressure fluids for engines of low power output.

Answer Your considerations are correct.

Question - Mr Macchi, Politecnico di Milano/CNPM : We made a study concerning ORC waste heat recovery from Diesel engines for a large Italian Firm. The first requirement was to adopt a safe and not flammable fluid in the recovery engine. I see that you selected Fluorinol and Toluene, which are flammable and toxic fluids, for ship applications. Do you think that this choice would be accepted by naval architects ?

Answer Up to now we have not made out choice, but we have considered some particular criteria from the thermodynamical point of view. The problems of safety must be considered and depend on the type of the ship and on the application.

Question - Mr Gaia, Turboden/CNPM : Why did you not consider Naval Diesel Engines larger than 2500 kW ? Actually an investigation is being made for heat recovery in a 10 000 or 15 000 HP ship by Italcantieri, Italy.

Answer We believe that the waste heat from these engines can be used by Cycle systems working with water.

MEDIUM TEMPERATURE 100 KW ORC ENGINE FOR TOTAL ENERGY SYSTEMS

Authors : G. ANGELINO, M. GAIA, E. MACCHI,

Contract number : EE-B-1-131-I

Duration : 36 months 1-1-80 - 31-12-82

Head of project : FERRANDO GAIA, TURBODEN s.r.l.

Contractor : TURBODEN s.r.l.

Address : Via Circo, 1
20123 MILANO, Italy

Summary

An organic working fluid heat recovery engine of 100 KW rating was designed which will provide a power output much higher (about twice as much) than achievable with a standard steam turbine, while applying a substantial heat flow at an usable temperature.
The following assumptions were made for the design point of the ORC engine:

- speed of revolution 3000 rpm
- evaporation temperature 150 to 200°C
- condensation temperature larger than 80°C

The latter assumption is due to the requirement of utilizing the heat released at the condenser for space heating or other thermal uses. O-dichlorobenzene was selected as working fluid, being well adapted for a 3000 rpm turbine due to its large molecular mass and low condensation pressure. A global efficiency (electric to thermal power ratio) in excess of 13% is expected, with a supersonic two stage axial turbine.
The turbine is fitted directly on the generator shaft; two subsequent test periods are planned, in the first one the turbine will revolve on standard ball bearings, with a mechanical seal to keep the shaft tight. In the second one the generator is separated by a casing and is cooled by the working fluid.
The comparatively large diameter turbine asks for the development of a fabrication technique compatible with the low final cost of the engine. A number of promising alternatives were actually tested.
Though the contract definition with EEC was greatly posponed, the research program is expected to be completed within October 1983.

1.1 Introduction

The Organic Rankine Cycle(ORC) engines are the most efficient converters of heat into electricity when the power level and temperatures involved are moderate (up to at least 300°C and 500 kW).

Hence, the use of ORC engines should be considered in many cases of thermal energy cascading applications.

The main objectives which led to the specifications for the present contract were:

a) The heat leaving the condenser should be at a temperature high enough for space heating or other low temperature uses during the Winter. In fact also a limited use of the condenser heat greatly improves the pay-back time of an ORC recuperator by improving the yield of the system and by reducing the cost of heat release in Summer months.

b) The power level should be high enough to justify the installation costs. These are necessarily high, each recovery case being different.

c) The efficiency of the thermodynamic cycle and of the expander should be kept as high as possible in order to obtain a favourable ratio of actual efficiency to maximum theorical efficiency, with moderate size heat exchangers. In particular the primary heat exchanger (exhaust fluid/ORC working fluid) is often an expensive and short lived component.

The aim of the present research is the design, construction and testing of an ORC engine having the following specifications:

- 100 kW electric power
- about 170°C evaporation temperature
- about 80°C condensation temperature

1.2 Description of the Plant

1.2.1 Rationale for shaft speed and working fluid selection

A turbine speed of revolution of 3000 rpm was assumed a priori. So low an angular velocity can be obtained in an ORC engine mainly by adopting a low condensation pressure, high molecular mass fluid, yielding a large diameter turbine. In the case of waste heat recovery at the selected rating, the disadvantage of higher turbine cost is largely overcome by the suppression of the speed reducing gear. A 3000 rpm system can be designed as a two bearing system with the generator rotor between the bearings and overhang turbine, giving a very simple and reliable mechanical arrangement. Moreover the overall efficiency is increased some 2% to 4% due to the suppression of the gears, the additional bearings and the couplings.

The auxiliary equipment complexity cost and energy consumption are also reduced by the suppression of the lubrication circuits and of the related safety devices. The shaft sealing at 3000 rpm can be solved by a mechanical seal with a low peripheral speed (12 to 15 m/s); a machine completely enclosed free of any rotating seal can also be designed by letting the generator and pump rotor run in the working fluid.

The main requirements for a suitable working fluid for the present application pertain to:

1) - thermodynamics

Good cycle efficiency is obtained from fluids having a critical temperature mugh higher than the temperature level of the heat source. A low liquid specific heat in comparison to the heat of vaporization (i.e. a working fluid molecule as simple as possible) is desirable for optimum efficiency.

2) - fluid - dynamics
The decision to employ a turbine expander poses additional contraints on:
- working fluid volume flow. Volume flow at condenser inlet must be consistent with the desired turbine rotating velocity. In particular, for a 3000 rpm turbine, a large volume flow is requested which, for a 100 kW power level, is achievable only at low condensing pressures (in the vicinity of the condensing pressures of ordinary steam plants).
- working fluid enthalpy drop. A conservative turbine mechanical design and reasonable wheel dimensions are favoured by a small enthalpy drop which is basically assured by a large working fluid molecular weight.

3) - thermal stability
The working fluid should be thermally stable at the maximum forseable heat exchanger inner wall temperature, which is somewhat higher than the top thermodynamic cycle temperature. Heat source characteristics, fluid transport properties and heat exchanger design determine the actual top temperature acting on the fluid. A thermal stability level compatible with the top heat source temperature would greatly simplify heat exchanger design and engine regulation philosophy.

4) - general technical acceptability. The level of technical acceptability of the working fluid with respect to toxicity, corrosion characteristics, flammability, cost etc. must be comparable with that of organic fluids widely used since decades in thermal equipment (such as refrigerants or heat carriers).

Chloro-fluoro substituted hydrocarbons (Freon-type) are unsuited for the present application lacking the requested level of thermal stability (they cannot be used above approximately 150°C) and giving rise to insufficient turbine volume flows for a 3000 rpm expander.

Fully fluorinated hydrocarbons (marketed as Flutec, Fluorinert etc. by different manufacturers) exhibit marginally acceptable thermodynamic properties (having excessively large molar liquid specific heats) and are probably too expensive for a power recovery application.

Hydrocarbon-based heat carriers do not have, in general, a sufficient vapour pressure at low (condensation) temperature to be used in a thermodynamic cycle. Among the few exceptions, o-dichlorobenzene (marketed, for instance, by DuPont de Nemours as Dowtherm E) seems to possess a good mix of properties to be proposed for the present application.

It was safely used in the past at temperatures in excess of 250°C at which its vapour pressure is substantially higher than atmospheric pressure. In the present case, it was decided to limit the top cycle temperature to 173°C so that everywhere in the working fluid loop, internal pressure is lower than that of the atmosphere. This should minimize working fluid release to the ambient.

Toxicity and flammability levels are reported to be moderate (TVA is presently 50 ppm). Thermodynamic and fluid dynamic properties are consistent with the requirements recalled in 1) and 2). Heat transfer wall temperatures up to 260°C should not degrade the working fluid.

1.2.2 Apparatus and test program

The plant comprehends the engine, the heat supply, the cooling system and the measuring instruments, according to fig. 1.

It is being erected at the Turboden test facility near Milano, in order to carry out the performance tests of the machine.

An engagement is also being defined for the installation and further testing of the plant using the heat recovered from a ceramic tile furnace in Sassuolo, Italy.

The primary energy source for the performance tests is a heat transfer oil flow, heated in a fossil fuel boiler. A modulated flame burner allows a wide range of primary thermal power. The power output of the engine is measured at the generator bus bars; the mechanical power at the generator shaft can be evaluated, the generator performance curves being certified for a wide range of power outputs. The main temperatures and the heat transfer oil and water flows are measured; the ORC working fluid flow is not measured, being easy to estimate the flow with good accuracy in the supersonic nozzles from the upstream temperature and pressure.

The performance tests (1st stage) are aimed to obtain the power output and the efficiency (electric power vs heat input) curves as a function of oil and cooling water temperatures. The influence of the incondensible gas content on the heat exchangers performance will also be checked, by introducing known amounts of gas. This information is relevant to the possibility of eliminating the vacuum pump.

More tests on the machine will include:

- measures of the leak rate of air into the system before and after thermal cycling of the heat exchangers,
- functional tests for the control and safety equipment
- performance tests with the alternator enclosed in a shell and cooled by the working fluid (2nd stage tests).

The following steps were performed, though the contract definition was largely delayed (March 1982):

- Final definition of the thermodynamic cycle
- definition of the design parameters of the main components
- detailed plant design
- the issuing of the purchase orders for the main components (heat exchangers, turbine parts, alternator, feed pump, test loop heater and cooling tower).

The construction of the turbine blades was also initiated.

The time schedule for the following work is given in fig. 2 the fullfillment of the contract being planned within October 1983.

1.3 Results

1.3.1 Thermodynamic cycle

As said before, an evaporation pressure of about 0.9 bar was adopted, in order to ensure that all components in the hot side of the cycle operate at a pressure below the atmospheric pressure. This was in fact considered desirable from the safety point of view; allowing a very quick detection of leaks and ensuring the shutdown of the plant before a dangerous leak could evolve.

The following options were considered (fig. 3):

- simple, non regenerative cycles
- cycles with regeneration downstream of the turbine
- combined cycles with an extraction heater and regenerator downstream of the turbine.

Adding a regenerator, with a 15°C minimum temperature difference, improves the performance by about 4%, while the improvement with an added optimized extraction heater is even smaller.

Though the gain of efficiency associated with regeneration is small, the regenerator was included in the design. In fact it provides a de-superheating of the fluid which is necessary for a good performance of the condenser and can be designed to produce a very limited pressure drop. The design includes also a spray nozzle system to test spray de-superheating, both with and without the regenerator in operation.

The main data for the adopted cycle are given in fig. 4.

1.3.2 Aerothermodynamic turbine design

A two stage, axial-flow turbine was selected. Other possible configurations would yield either poor and difficult to predict performance(single stage radial or axial), or unjustifiably complex and costly solutions (multistage turbine).

Considerations related to simplicity of mounting suggested the following rather unconventional requirement:

- first stage outer radius ≤ second stage stator inner radius. An other requirement was the design speed of revolution, fixed to 3030 rpm (1% over synchronism), since the turbine is directly mounted on the generator shaft.

The design procedure was mostly based on:

- a computer-aided optimization technique, for the selection of the main design variables. A description of the method can be found in [1] and [2], where assumed loss correlations are discussed;
- the adoption of blade profiles successfully used by Turboden in previous machines.

An iterative procedure was adopted, to obtain a solution from the computer program compatible with available blade profiles.

An overall adiabatic efficiency of 82,8% was predicted for the final turbine configuration, by assuming a kinetic energy recovery factor (ϕ_E) of 50% downstream of the second stage. The efficiency is still over 81%, by not considering this recovery.

The two stages operate with similar enthalpy drops and pressure ratios. A large number of blades is required due to the relatively low adopted value of specific speed.

As shown by the loss analysis given in tab. I greater losses occur in the low aspect ratio first stage than on the second one, which operates at higher peripheral speed , and therefore at higher degrees of reaction.

The cross section of the turbine is given in fig. 5.

	1 stage	2 stage
Stator	4.27	3.16
Rotor	3.17	1.69
Leakages	1.01	1.05
Kinetic energy	1.25	1.56
	9.70	7.46

Tab. I Predicted loss breakdown in the turbine. Losses are expressed as percentage of the isentropic enthalpy drop across the turbine.

1.3.3 Electric generator selection

An asyncronous generator was adopted due to the following reasons:

a) the engine was designed to be used for heat recovery in industrial environment, therefore to be connected to a powerful electric supply which can give the necessary excitation power at fixed frequency.
b) The parallel operation of a relatively small size asyncronous generator is simpler and safer. Following a grid failure the generated tension rapidly falls to zero.
c) The squirrel cage rotor of an asyncronous generator has lower cooling requirements than the windings and the solid state devices of a brushless syncronous alternator.
d) The efficiency at peak power for an asyncronous generator is good, being often higher then the one of a synchronous alternator. The efficiency falls at part load operation, however the normal operation of a heat recovery engine should be at high load, in order to allow a short pay-back time.
e) The cost of an asyncronous generator is lower, or, alternatively, a more efficient, better protected generator can be bought for the same price. Also, it could be easily re-used as a motor after the end of the engine operating life.

The layout for the engine is given in fig. 6.

1.3.4 Manufacturing

A low cost for the vapour expander in spite of its large dimensions is a primary requirement for the cost effectiveness of the machine. Hence, milling of the blades from solid forged wheels, i.e. the standard Turboden technique, was abandoned also for the prototype engine.

Experiments of construction of single blades were done, as well as market inquiries. The results are summerized in table II, with reference to light allow blading, mounted on a high strength steel disc. The experiments of hot press powder metallurgy, which was never up to now considered for ORC engine blades, gave interesting results.

METHOD	SET UP COST	BLADE COST	TIME DELAY	PROFILE ACCURACY
EXTRUSION	MODERATE	LOW	LARGE	LOW
COLD ROLLING	HIGH	LOW	VERY LARGE	GOOD
FORGING	HIGH	LOW	LARGE	LOW
FORGING & COINING	HIGH	MODERATE	LARGE	FAIR
INVESTMENT CASTING	MODERATE	HIGH	MODERATE	FAIR
CNC MILLING	LOW	MODERATE	MODERATE	EXCELLENT
SINTERING	MODERATE	MODERATE	MODERATE	FAIR

Tab.II Results of experimental work and inquiries for the turbine blades.

1.4 Comments

It is of interest to compare the predicted performance of the proposed ORC engine with results achievable by resorting to a more conventional steam cycle, having a commercially available turbine as expander.

The comparison is carried out by assuming a saturated steam cycle having evaporation and condensation temperatures (173.6 and 80°C respectively) and thermal input (744 kW) equal to design values stipulated for the ORC engine.

It can be seen that steam and dichlorobenzene yield cycles having similar thermodynamic qualities, so that variations on mechanical/electrical output achievable by the two working fluids are strictly related to different turbine efficiencies. The ORC turbine efficiency was computed by means of a theoretical loss prediction method the validity of which was confirmed by experience on several ORC of similar design characteristics. Steam turbine efficiency were instead computed according to the performance curves given by a large U.S. manufacturer for two single-stage models, having a 350 and 500 mm. mean diameter and 6000 and 12000 rpm maximum speed respectively.

The curves of fig. 7 demonstrate that very low efficiency values are obtained for both steam turbines; their poor performance is mostly caused by too low blade speed coefficients (350 mm. turbine) and high ventilation losses (500 mm. turbine). As a consequence, the resulting power output obtainable with a steam cycle is much lower that the ORC one, as shown in fig. 8. Low efficiency are obtained for the steam turbines also for low condensation temperatures (dotted curves in fig. 7), since the decrease of ventilation losses are more than counterbalanced by the loss increase related to larger enthalpy drop. Only at much larger power output the high speed turbine efficiency exhibits marked improvements (see curves for N = 1000 kW in fig. 7), anyway reaching values well below the one predicted for ORC.

Obviously, the above results hold only for commercially available steam turbines, since it would not be impossible to design a high efficiency turbine in the low power output range; however, it would be necessary to resort to a multi-stage turbine, having high peripheral speed and consequently high mechanical stresses and running at very high speed of revolution. All these requirements would make this machine costly and hardly attractive for an industrial application.

The use of an Organic Rankine Cycle engine instead of a steam cycle makes it possible to avoid these problems, by selecting a working fluid with more desirable characteristics for the turbine designer: low enthalpy drop and appropriate volume flow rate.

2. Conclusions

The proposed Organic Rankine Cycle yields a mechanical/electrical power output almost double than the one achievable by a more convential solution adopting a commercially available steam turbine. This feature, typical for plants having outputs in the range of few hundred kilowatts, makes ORC a valuable energy-saving solution both for industrial waste heat recovery or for cogenerating plants which discharge heat at temperatures well above the ones required by thermal utilizers, such is the case with Diesel engines and gas turbines.

The aim of the present research is to demonstrate that it is possible to design and commercialize a ORC unit of this size at reasonable costs.

By selecting a low pressure working fluid, it can be designed an ermetically sealed unit running at 3000 rpm with a very conservative mechanical design: potentially, this unit should have availability characteristics to be utilized in industrial environment.

3. References

1 E. Macchi, A. Perdichizzi, Efficiency Prediction for Axial Flow Turbines operating with Nonconventional Fluids, Journal of Engineering for Power, vol. 103 Oct. 1981.

2 G. Lozza, E. Macchi, A. Perdichizzi. On the Influence of the Number of Stages on the Efficiency of Axial Flow Turbines ASME Paper No. 82-GT-43, New York 1982.

DISCUSSION

Question - Mr Zegers, CEC : ORC engines have been on the market for several years. Is the type you are investigating on the market ? What are the possibilities for application in industry ?

Answer ORC engines of a similar power are constructed by some manufacturers (e.g. Ormat of Israel). But we feel that our design has particular features which makes it very promising i.e. high condensation temperature associated with high efficiency, thermally stable fluid, extreme simplicity.

Comment - Mr Wood, Cranfield Institute : I agree that low power i.e. $\leq$100 kW shaft devices are likely to be easier to integrate into processes due to the less influence they have on the complexity of the plant. The market for such devices will be more diffuse but greater in volume than high power ORCS.

Question - Dr Hussein, Cranfield Institute : 1. Are the blades made of Al-28 ?
2. How about temperature sources of 400-500° C ?
I believe that these high temperature sources could be used directly as a pre-heater for the industrial process.

Answer 1. The blades are machined from light alloy bars by CNC milling.

2. Each recovery case has different features. As a general rule the bottoming cycle has to be considered after all the possible thermal energy utilisations have been evaluated.
I think that the proposed engine, which in a way is more an intermediate cycle than a bottoming cycle, can improve the overall thermodynamic performance of many processes.

Question - Dr Pilavachi, CEC : In your system you have low pressures. As a result do you find that the heat exchangers would be very large ?

Answer The phase change heat transfer coefficients are not largely influenced by the operating pressures.

Question - Mr Hopmann, MBB : 1. What is the height of the turbine blades ?
2. Is the cost of erosion milling being considered ?
3. What is the material of the disk and the blades ?

Answer 1. The height of larger blade is some 60 mm.

2. No precise evaluation of blade cost is available at this time.

3. The disc is made of 4340 steel. The blades are made of a wrought light alloy, mainly used for aircraft parts.

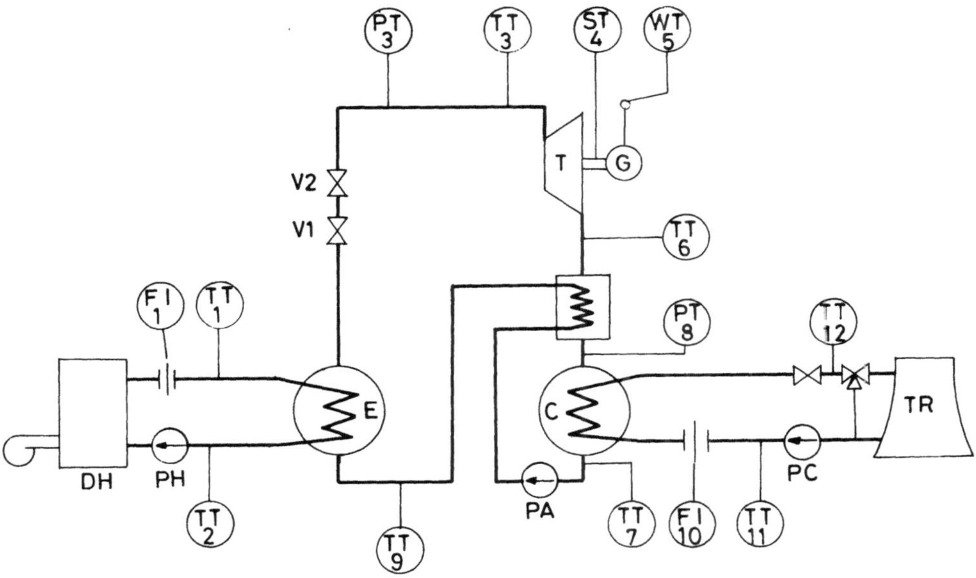

Fig. 1 TEST APPARATUS

OH	Oil Heater	T	Turbine	PC	Cooling Water Pump
PH	Heat Transfer Oil Pump	G	Electric Regenerator	TR	Cooling Tower
E	ORC Evaporator	R	ORC Regenerator		
V1	ORC Control Valve	C	ORC Condenser		
V2	ORC Control Valve	PA	ORC Feed Pump		

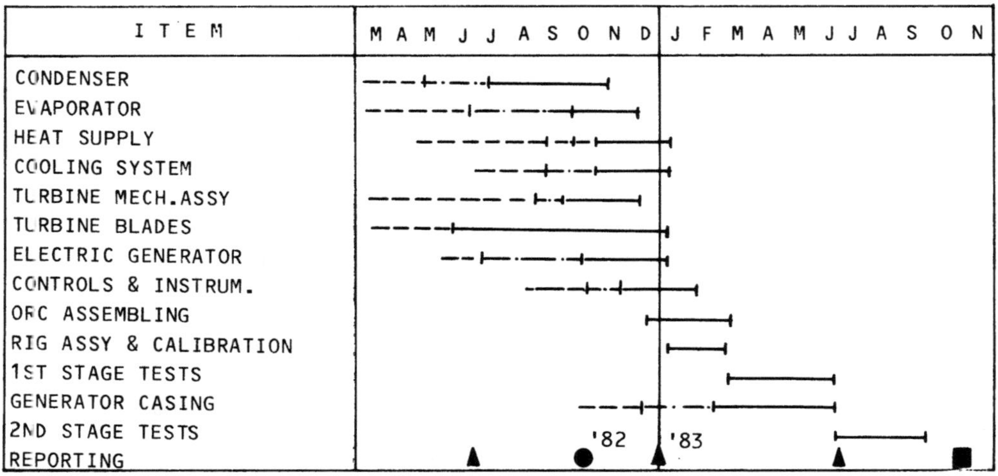

Fig. 2 TIME SCHEDULE

DESIGN & SPECS FOR PURCHASE
CALL FOR BIDS
PURCHASE OR CONSTRUCTION
INTERMEDIATE REPORT
FINAL REPORT

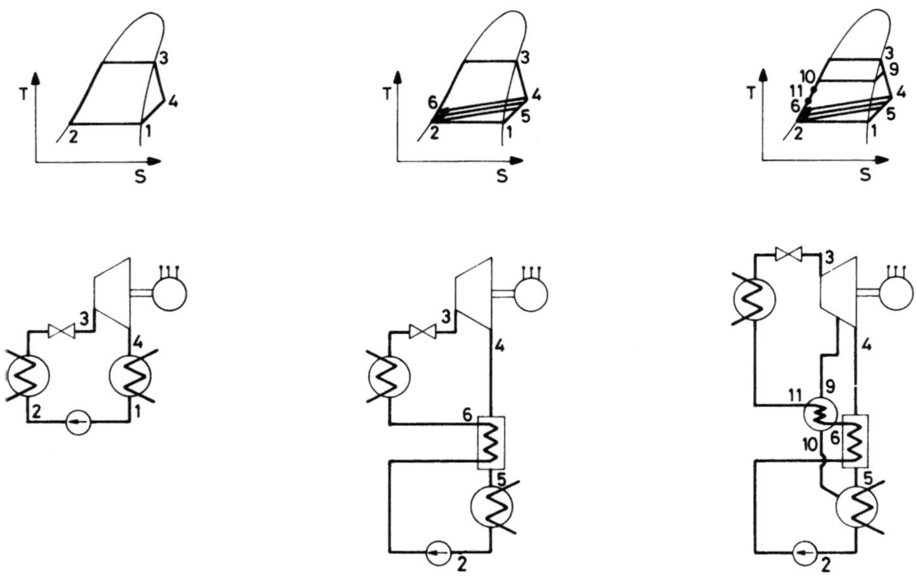

Fig. 3 CYCLE CONFIGURATION CONSIDERED

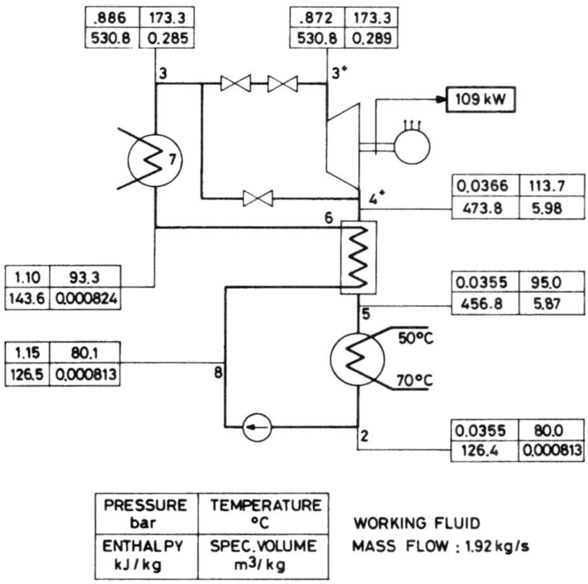

Fig. 4 MAIN POINTS FOR THE ADOPTED CYCLE

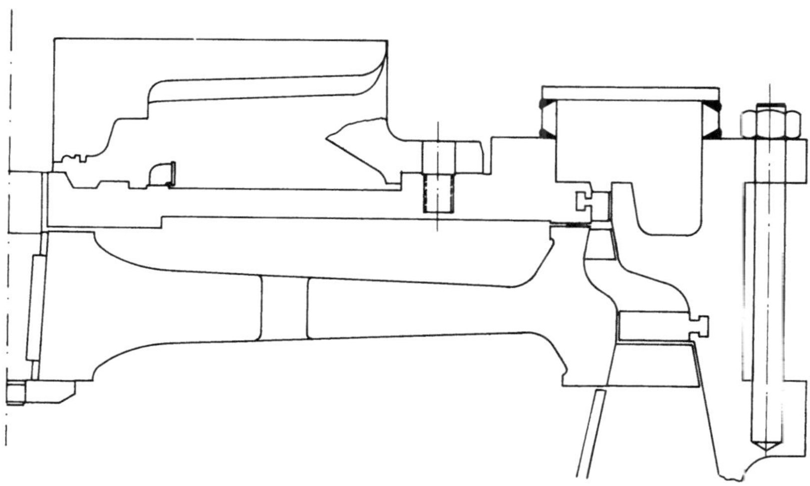

Fig. 5 SIMPLIFIED TURBINE CROSS SECTION

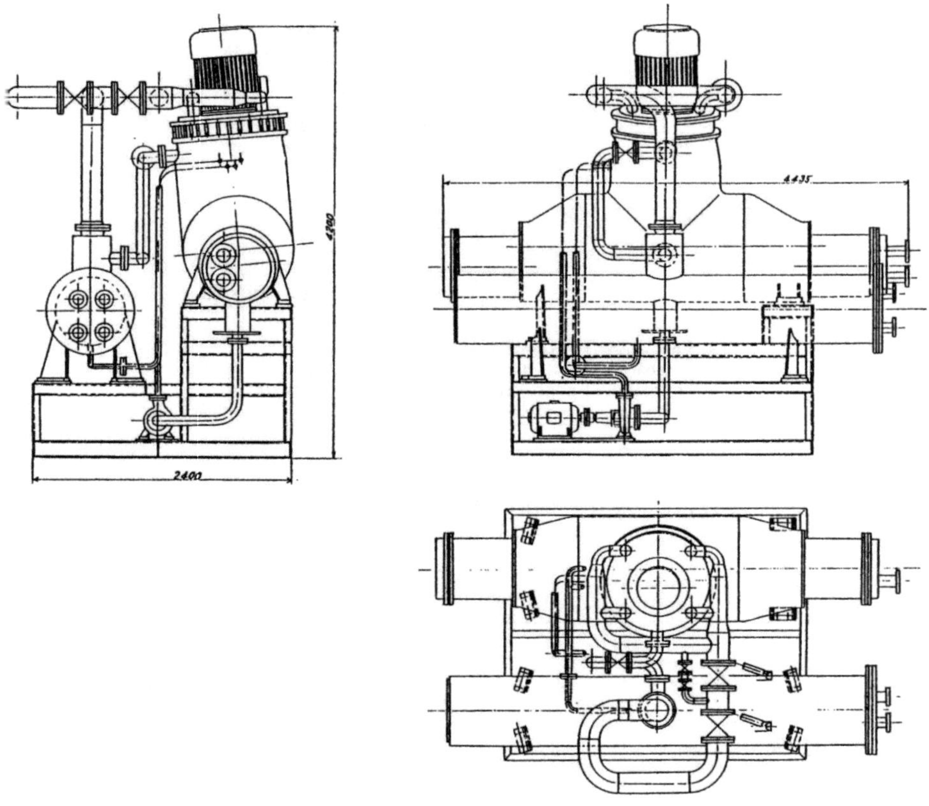

Fig. 6

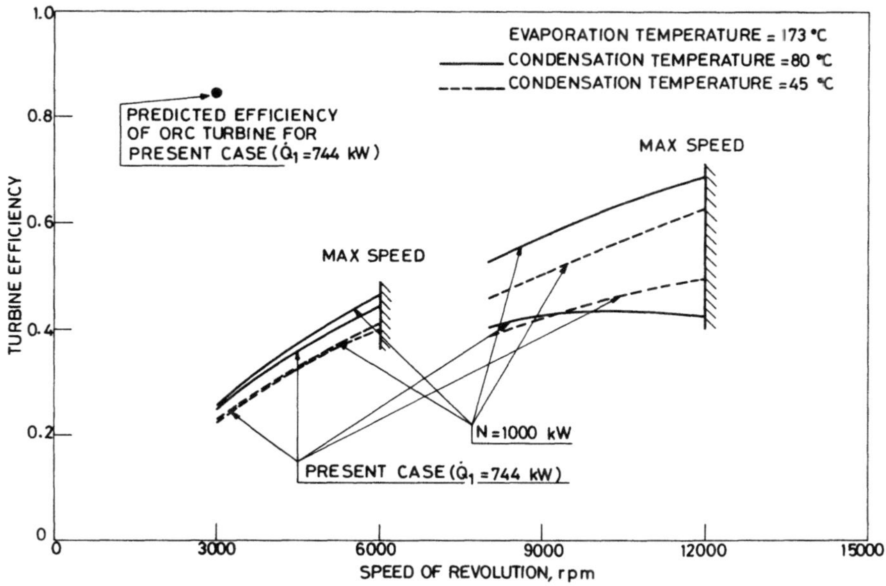

EVAPORATION TEMPERATURE = 173 °C
CONDENSATION TEMPERATURE = 80 °C
CONDENSATION TEMPERATURE = 45 °C
PREDICTED EFFICIENCY OF ORC TURBINE FOR PRESENT CASE ($\dot{Q}_1$ = 744 kW)
MAX SPEED
MAX SPEED
N = 1000 kW
PRESENT CASE ($\dot{Q}_1$ = 744 kW)
TURBINE EFFICIENCY
SPEED OF REVOLUTION, rpm

Fig. 7

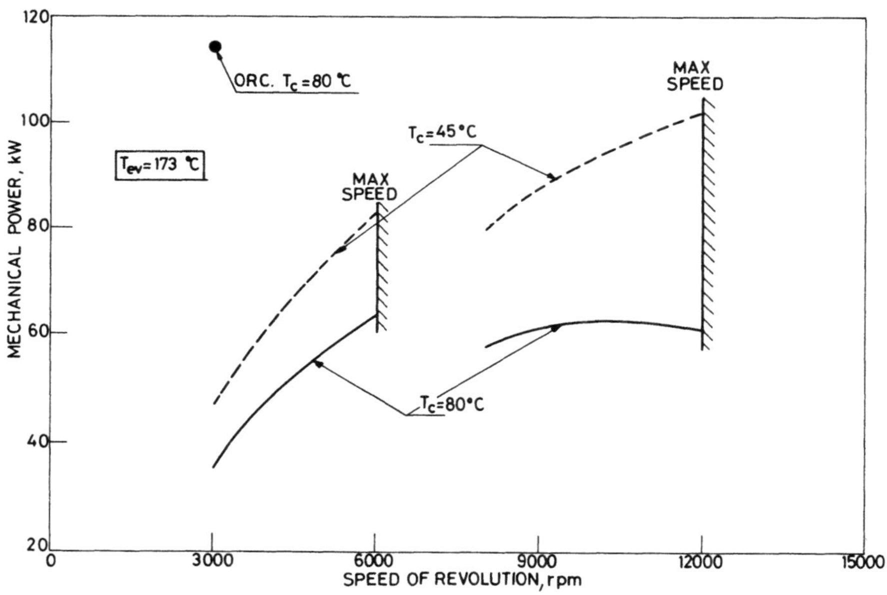

ORC. T_c = 80 °C
T_{ev} = 173 °C
T_c = 45 °C
MAX SPEED
MAX SPEED
T_c = 80 °C
MECHANICAL POWER, kW
SPEED OF REVOLUTION, rpm

Fig. 8

OPTIMISATION OF A MULTI-VANE EXPANDER AS THE PRIME MOVER IN AN ORGANIC RANKINE CYCLE

Authors : P.W. O'CALLAGHAN, MOHEY HUSSEIN, S.D. PROBERT, M.A. BELL and R.M. PATEL

Contract number : E/B1/121/GB

Duration : 24 months: 1 November 1980 - 31 October 1982

Head of project : DR. P.W. O'CALLAGHAN, Reader in Energy Management, Cranfield Institute of Technology

Contractor : Cranfield Institute of Technology

Address : Cranfield, Bedfordshire, MK43 OAL, United Kingdom

Summary

This paper describes the research and development activities conducted by the Low-Grade Energy Group at Cranfield Institute of Technology in conjunction with Denco Air Ltd., Hereford, regarding the performance of multivane expanders as prime movers in organic Rankine cycles. Intensive, fully-descriptive, computer-aided design studies have been completed and have led to the development of robust, low-cost expansion devices with unique characteristics suitable for use in low-grade energy systems using:-

industrial waste heat recovery; geothermal energy; solar energy; ocean thermal energy potential differences; bottoming cycles for power generation.

The main features of the machines are:-

flat operating efficiency characteristics over a wide range of operating conditions; low speeds (∿ 3000 rpm); ability to tolerate ingress of liquids and wet vapours; easy controllability; high power-to-weight ratio low maintenance requirements; few easily replaced moving parts; low lubrication requirements.

A further short design phase is about to commence during which expanders with greater efficiencies are expected to evolve. A complete suite of Fortran and Basic computer programs have been established during the theoretical study covering:-

working fluid properties for water, R11, R12, R13, R22, R502, R113, R114, and any halocarbon refrigerant for which a fully-specified equation of state exists; the theoretical performance of organic Rankine cycles; geometrical and dynamic models describing the expander, and its irreversibilities - friction, leakage and thermal losses.

A flexible experimental test facility has been built and commissioned to test various prime movers and Rankine-cycle systems.

Extensive practical testing of two types of multi-vane expander has been accomplished.

1. Introduction

A multi-vane expander (MVE) is a positive-displacement expansion device, whose principle of operation is identical to that of a piston or screw expander.

The use of an efficient MVE as the prime-mover in an organic Rankine cycle (ORC) offers many advantages, foremost of which are:-

a) simple design and rugged construction;
b) few moving parts, e.g. no valves;
c) speed directly compatible with pumps, generators, fans, compressors, etc., i.e. no gearbox is involved;
d) high reliability and low maintenance;
e) quiet in operation;
f) can expand into the wet region of a working fluid without adverse consequences;
g) relatively flat efficiency curve and good performance through a wide range of operating conditions.

MVE's used as compressed air motors are very inefficient - 20→30% being typical (1). Multi-vane compressors tend not to be as efficient as their reciprocating piston counterparts, and the aim of this research programme has been to identify, quantify and minimise those features of a MVE which lead to reduced efficiency.

The task of optimising the design of a MVE has proceeded simultaneously along both theoretical and practical routes. The MVE cannot be studied in isolation: therefore consideration was first given to possible working fluids of the Rankine cycle (2).

Analytical models of the MVE geometry and mechanical (friction) characteristics were derived for three different designs of MVE in order to ascertain the inherent performance limitations of each. An analytical model for the internal leakage was also derived. Each of the analytical models was compared with results from practical tests.

From the combined experience of the theoretical and practical work, new multi-vane expanders were designed and tested. Finally an optimisation procedure was established and two examples are described.

2. Selection of the Rankine Cycle Working Fluid (Ref. 2)

The correct choice of the working fluid in the Rankine cycle is a critical factor, because it is the "lifeblood" of the system. Its effect upon the design and performance of individual cycle components and hence upon the overall performance is paramount. Consequently the cost of the system and the economics of its particular application are affected by the choice of the working fluid.

Certainly, the ideal working fluid should have desirable physical, chemical, transport, physiological, handling and thermodynamic properties. A candidate fluid should be chemically stable at the maximum expected cycle temperature and when in contact with the engine materials, lubricating oil, and unavoidable contaminants. Preferably, it should also be non-corrosive, non-flammable, non-toxic, and cheap.

In utilizing an organic working fluid, several constraints have to be imposed so that a satisfactory long-life operation (i.e. up to 10 years) is achieved. These include the thermal stability of the working fluid in the presence of both the lubricating oil and the metals of the system components.

Table 1 summarises the characteristics of candidate working fluids.

Five of these fluids were selected as suitable for the present application, involving as it does temperatures as high as 200°C (Table 2).

Refrigerants R11 and R113 have already been tested for stability (e.g. Figure 1) and compatability with various synthetic and mineral oils. An upper bound of 143°C has been identified.

3. Design Studies

Three designs were analysed to make comparisons: a Conventional MVE; a Lobe MVE and an Axial MVE (Figure 2).

There are five fundamental problems associated with the development of an elevated temperature, high-pressure, multi-vane, vapour expander if both durability and high efficiency are to be achieved:-

proper material combinations including vane/stator, vane/rotor and vane/end plates; internal leakage; internal friction; breathing of the expander; choice of the working fluid.

3.1 The Analysis

A mechanical (or friction) analysis of the forces acting on a vane was performed for each of the above designs of MVE. The main assumptions were:- negligible vane thickness and constant rotational speed.

The major contribution to mechanical friction in a MVE arises from the sliding contact between the vane and the rubbing surface. Other minor mechanical losses from bearings, seals, viscous shear, etc., termed "parasitic friction losses" also occur, but for the preliminary analyses these were assumed to be negligible.

From the mechanical analysis developed it is possible to predict:-

the instantaneous friction torque arising from a single vane; the total mean frictional torque; and the zones where the vanes lose contact with the rubbing surface.

Figure 3 shows typical plots of the instantaneous frictional torque for the conventional MVE and the lobe MVE under investigation. The rubbing surface profile in each case was based on cosine expansion and contraction sections.

The major frictional torque occurs during the contraction section (between θ_3 and θ_4). Whereas with the conventional MVE, loss of vane contact tends to occur in the expansion sector (as the speed falls), with the lobe expander, it tends to occur in the inlet sector (as the speed rises).

Figure 4 summarises the results of the design study, where the theoretical mechanical efficiency for each design is plotted against speed.

The lobe MVE demonstrates the possibility of a high and virtually constant mechanical efficiency, whereas the mechanical efficiency of the conventional MVE falls with speed owing to the increasing centrifugal forces on the vanes. The axial MVE appears inferior to the lobe MVE. Because there were likely to be additional problems with vane sealing against a helical surface and some uncertainty as to the validity of the assumptions in the force analysis (mainly the centrifugal loads on the vane ends and force distribution along the vane width), it was rejected as a viable candidate for further development.

It should also be noted that the conventional MVE must be operated somewhat above a critical speed in order to avoid excessive internal leakage with the consequent loss in efficiency. The lobe MVE may be operated above an upper critical speed because the loss of vane contact does not occur (initially at least) in the expansion sector.

Because of the preliminary results of this study, it was decided to build a prototype lobe MVE and test it alongside a conventional MVE.

One-dimensional viscous flow theory together with some empirical data were applied to three major leakage paths across (i) the seal sector, (ii) the side face, and (iii) the vane ends.

Other minor leakage paths have also been identified, but these were considered to be at least an order of magnitude smaller than the above and therefore not considered further.

4. Technical Specifications of the Test Rig (Ref. 2) (Fig. 5)

Working fluid - Refrigerant R11, R113 or steam

Maximum mass flow rate - 0.27 kg/sec.

Evaporator

Shell-and-tube type heat exchanger

Two-pass tube side (16 tube - $\frac{3}{8}$" o/d)

R11 on shell side, steam on tube side

Disengagement space with central vapour take-off

Capacity 25kW to 55kW

Temperature 70°C to 150°C

Pressure 4×10^5 Nm^{-2} to 21×10^5 Nm^{-2}

Condenser

Shell-and-tube type heat exchanger

Three-pass tube side (12 tube - ϕ 0.512"/ϕ 0.505")

R11 on shell side, water on tube side

Disengagement space, vapour introduced through sparge pipe

Capacity 20kW to 50 kW

Temperature 30°C

Pressure 1.25×10^5 Nm^{-2}

Expander

Multi-vane expanders

Design shaft power up to 5kW

Cooling Water

Mass flow rate 0.9 kg/sec to 1.2 kg/sec

Temperature 15°C

Heating Source

High pressure steam delivered from an electrode steam boiler

Capacity 65kW

Capacity range 10% to 100%

Steam flow rate 0.03 kg/sec at 100°C

Maximum steam pressure 10.34×10^5 Nm^{-2}

Maximum steam temperature 180°C

Liquid Feed Pump

Positive displacement multi-vane type campatible with the multi-vane expander

Maximum mass flow rate 0.27 kg/sec

Externally driven by electric motor

Control System

A level controller situated in the condenser controls the liquid level by regulating the feed pump by-pass. The by-pass valve is pneumatically controlled by the level controller.

System Safety Control System

- Air failure, a signal via the level controller, causes by-pass valve to open.
- High pressure switch on evaporator to trip out feed pump and steam boiler.
- High pressure switch on the condenser to trip our feed pump and steam boiler.
- Low pressure switch on the condenser, with manual over-ride, to trip out boiler and feed pump in the event of total refrigerant loss.
- Pressure relief valve is mounted on the evaporator,which opens to the atmosphere in case of the occurrence of high pressure in the evaporator.

The following parameters were measured:- torque, speed; inlet and outlet pressures/temperatures; volume flow; instantaneous cell pressure.

The accurate measurement of the latter, via a rotor mounted miniature pressure transducer with commutated output is vital to separate indicated and mechanical efficiencies. Considerable difficulties were encountered in the calibration and use of this device. Figure 6 shows typical data outputs from which qualitative conclusions only can be drawn. Further work in this area is vital in order to achieve continual design improvements.

5. Results

Refrigerant R11 was chosen for initial testing. Its performance appeared satisfactory, but it was known that progressive degradation of R11, especially when in contact with lubricating oils, occurred at temperatures above 105°C (3). However no evidence of significant degradation was found during the present test period, the bulk fluid temperatures never exceeding 120°C.

The ready availability of steam led to its use during a second phase of testing, where the MVE performance with wet expansion was evaluated. Work currently in progress employs R113 for testing a conventional MVE, and compressed air is being used for initial testing of a lobe MVE.

The experimental results presented in Figure 7 describe:-

a) Testing of the reversed multi-vane compressor using R11.
b) Testing of the reversed multi-vane compressor using steam.
c) Testing of the prototype lobe MVE using compressed air.

Prior to presenting these results however, it is necessary to specifically define the various efficiency factors employed.

5.1 Method I

The Efficiency Factors of a MVE

The efficiency of a practical MVE as compared with an ideal MVE can be best described with reference to a pressure-volume diagram (see Figure 8)

With reference to Figure 8:

P_I = inlet pressure
P_E = exhaust pressure
V_C = clearance volume
V_I = geometric inlet cell volume
V_E = geometric exhaust cell volume
V_{opt} = cell volume for the optimum volume ratio

The Effective Volume Ratio (VR_{eff})

The geometric volume ratio is given by V_E/V_I, but in practice the clearance volume and internal leakage (defined below) change the volume ratio as far as the working fluid expansion is concerned. The effective volume ratio is therefore defined as:

$$VR_{eff} = \frac{V_E + V_C}{(1 + C_I)\, V_I + V_C}$$

The Internal Leakage Coefficient (C_I)

C_I is a measure of the amount of working fluid which leaks into other cells tending to increase the amount of P-V work as compared with an ideal expander.

This is defined as the proportion by which V_I must be increased in order that the calculated indicated work equals the measured indicated work. Thus:

Measured indicated work = Area bdegi

or

$$W_{indic} = P_I\left(\left[1+C_I\right]V_I+V_C\right)\left\{1+\frac{VR_{eff}^{\,1-n}-1}{1-n} - \frac{VR_{eff}}{\left(\frac{P_I}{P_E}\right)} - \frac{V_C}{\left(\left[1+C_I\right]\,V_I+V_C\right)}\left(1-\frac{1}{\left(\frac{P_I}{P_E}\right)}\right)\right\}$$

The Cut-Off Efficiency* (η cut-off)

The cut-off efficiency is the ratio of the measured indicated work as compared with the work of an ideal expander under the same conditions. Thus

$$\eta_{cut\text{-}off} = \frac{\text{Area bdegi}}{\text{Area adfj}}$$

It can be expressed as:

$$\eta_{cut\text{-}off} = \frac{1+\dfrac{VR_{eff}^{\,1-n}-1}{1-n} - \dfrac{VR_{eff}}{\left(P_I/P_E\right)} - \dfrac{V_C}{\left[1+C_I\right]\,V_I+V_C}\left\{1-\dfrac{1}{\left(P_I/P_E\right)}\right\}}{1+\dfrac{\left(P_I/P_E\right)^{\frac{1-n}{n}}-1}{1-n} - \left(P_I/P_E\right)^{\frac{1-n}{n}}}$$

The Mechanical Efficiency (η_{mech})

The mechanical efficiency is conventionally defined as:-

$$\eta_{mech} = \frac{\text{Measured shaft work}}{\text{Indicated work}}$$

The Parasitic Leakage Coefficient (γ_L)

This is a measure of the amount of working fluid which leaks from the inlet to the outlet without accomplishing any expansion work whatsoever.

From the PV diagram, the mass-flow which accomplishes work is given by:

$$\dot{m}_w = \left(\left[1+C_I\right]\,V_I+V_C\right)\rho_I\,NZ$$

where:

ρ_I = inlet fluid density;
N = number of revolutions/second; and
Z = number of cells (or vanes).

If $\dot{m}$ is the total measured mass flow then γ_L is defined as:-

$$\gamma_L = \frac{\dot{m}-\dot{m}_W}{\dot{m}}$$

The Expander Efficiency (η_{exp})

$$\eta_{exp} = \frac{\text{Measured shaft work}}{\text{Isentropic expansion work of the measured mass flow}}$$

Using the above definitions it can be readily shown that

$$\eta_{exp} = \eta_{cut\text{-}off}\;\eta_{mech}\;(1-\gamma_L)$$

---

* Also known as the indicated efficiency.

Results from Testing the Conventional MVE using R11

The test results using the reversed multi-vane compressor were disappointing in that all attempts to measure the indicated power failed. The two major reasons for this were:-

a) Faulty pressure transducers: these were purchased as sealed absolute pressure transducers, but were found to leak. They therefore could not register the pressure accurately and subsequent ingress of R11 eventually rendered them unserviceable.
b) Calibration difficulties arising from a) above, and also the difficulty in establishing bench marks (known pressures) from the pressure/angle trace.

In spite of the above problems the pressure/angle traces obtained were useful in identifying irreversibilities within the cycle (Ref. 2).

However, as a result, it was necessary to resort to an approximate method in order to estimate the efficiency parameters. The technique will not be elaborated here but basically it involved calculating the mechanical efficiency using assumed data in the mechanical analysis computer program, and combining this with the measurements. An estimate for C_I was found by dividing the measured torque by the calculated nett torque. The results are shown on Figure 7a.

Testing the Conventional MVE using Saturated Steam

The same approximate analysis as described above was used with steam. The results are shown in Figure 7b.

Testing the Lobe MVE using Compressed Air

With the lobe expander, the pressure transducer remains stationary and is inserted through the outer casing. Therefore it was possible to use one of the absolute pressure transducers, which could not be installed into the rotor of the conventional MVE. The analysis for the lobe MVE followed precisely the parameters given earlier. The results are presented in Figure 7c.

5.2 Method 2

The measured performance data and the different efficiencies involved in the measured performance characteristics can be described as follows:-

Measured data or efficiency	Units	Efficiency definition	Accuracy of measuring, or accuracy of calculation
• Speed	rpm	-	± 1rpm of the measured speed
• Torque	Nm	-	± 0.25%
• Power	kW	-	± 0.27%
• Mass Flow Rate	kg sec^{-1}	-	± 0.49%
• Leakage Eff. (η_L)	%	$1 - F_L$ $F_L = \frac{\text{leakage mass flow}}{\text{total measured mass flow}}$	≃ ± 0.49%

Measured data or efficiency	Units	Efficiency definition	Accuracy of measuring, or accuracy of calculation
• Rankine Cycle Eff. (η_R)	%	$\frac{\text{measured shaft power}}{\dot{m} \times (\Delta h) \text{ across the evaporator}}$ (excluding the effectiveness of the evaporator)	$\simeq \pm$ 0.93%
• Carnot Cycle Eff. (η_C)	%	$\frac{\text{max. absolute cycle temp} - \text{min. absolute cycle temp.}}{\text{max. absolute cycle temp.}}$	
• MVE's Isentropic brake Eff. (η_{is})	%	$\frac{\text{work done per unit mass of working fluid}}{\text{isentropic enthalpy change from the inlet manifold conditions to the exhaust manifold pressure}}$	$\simeq \pm$ 0.93%

Figure 7d presents the measured performance map constructed according to this approach. (Ref. 2)

6. The Organic Rankine Cycle Optimisation

The overall efficiency of the cycle may be expressed as

$$\eta_{OA} = (\eta_{mech}\,\eta_{cut\text{-}off} - \frac{fp}{\eta_{pump}})\,(1 - \gamma_L)\,\eta_{ideal\ cycle}$$

where fp = power fraction of the ideal cycle required to pump the liquid to the boiler;

η_{pump} = feed pump efficiency; and

$\eta_{ideal\ cycle}$ = ideal Rankine cycle efficiency

(Other parameters are as defined previously).

If constrained to fixed upper and lower pressures/temperatures, $\eta_{ideal\ cycle}$, fp, η_{pump} and γ_L can be considered invariant. An optimisation therefore consists of maximising the product of η_{mech} and $\eta_{cut\text{-}off}$.

Because $\eta_{cut\text{-}off}$ is a function of volume ratio VR, the MVE is optimised with respect to VR. A conventional MVE may change its volume ratio either with constant aspect ratio or constant rotor radius. The optimisation is possible because η_{mech} tends to fall as VR increases, and $\eta_{cut\text{-}off}$ rises to a peak and then falls. Figure 9 shows the results of optimising a 5kW conventional MVE using R11.

If the constraint of fixed upper and lower temperatures does not exist then the inlet pressure may be changed. An optimum then occurs because, as the inlet pressure rises, $\eta_{ideal\ cycle}$ rises and γ_L also rises (so that $(1-\gamma_L)$ falls). A complete optimisation involves optimising for VR at each pressure level. For the example given in Figure 10 (i.e. for the lobe expander) however, η_{mech} remained substantially constant irrespective of the pressure level. Therefore the optimisation consisted of maximising the product $(1-\gamma_L) \cdot \eta_{ideal\ cycle}$.

Figure 10 shows the results of optimising a 5 kW lobe MVE using steam for two orders of the internal clearance.

7. Discussion

7.1 Working Fluid

Only in recent years has interest grown in Rankine cycle working fluids other than water. Nevertheless, there is still considerable lack of working experience with many of the fluids of interest. Thermodynamic data for some fluids are either incomplete or difficult to obtain and information concerning thermal breakdown, catalytic effects of oils and metals and toxicity is frequently non-existent. In addition the cost of some of the less-common fluids is very high.

The selection of R11 and R113 was based on their thermodynamic suitability for low temperature Rankine cycles (100→150°C) together with their low cost and easy availability.

Saturated steam appears to have much to commend it. Water is of virtually zero cost, non-toxic, non-flammable, and non-corrosive (if oxygen is excluded). The use of steam gives high thermodynamic efficiencies and does not require a regenerator. In addition, steam boiler design and operation is a well-known and long-established technology.

7.2 The Mathematical Models and Analyses

Although the mathematical models enabled considerable insight into the qualitative aspects of the operation of MVE's (e.g. loss of vane contact), their use for accurate quantitative prediction was limited by the unavailability of accurate data. In the mechanical analysis a reliable value for the sliding coefficient of friction of the vane tip is needed. This can only be found from accurate measurements of the indicated work combined with a good estimate of the friction losses. Owing to the pressure transducer problems encountered this could not be achieved for the MVE.

For the leakage analysis, an accurate knowledge of the internal working clearances and the working fluid viscosity is required. The latter becomes very difficult to assess when the leaking medium is a mixture of gas, oil and/or condensed liquid.

In spite of these difficulties, providing reasonable assumptions are made, the mathematical models can predict the MVE performance with reasonable accuracy as is shown by the good agreement between the predicted and measured results for the experimental devices.

The major omission from the mathematical models arises from the inability to predict a value for the internal leakage co-efficient. While this is by no means impossible, it is likely to be a daunting task.

7.3 The Conventional MVE using R11

Extensive testing was carried out using R11 (Ref 2). An initial problem with an over-heating seal was experienced but required only minor modifications to correct. Lubrication was a major problem because oil injected into the MVE accumulated in the boiler. The problem was circumvented by lubricating with cooled oil-rich boiler liquid.

The pressure/angle traces indicated that both inlet throttling and loss of vane contact were occurring within the expander. Unfortunately the pressure/angle traces could not be used for accurate estimations of the indicated power because of calibration difficulties.

Predictions using the mathematical models indicate that with a refined geometry and close control of internal clearances, conventional MVE efficiences in the range 70-80% should be attainable.

7.4 The Conventional MVE using Steam

The MVE used with steam performed well, but its efficiency was much affected by the amount of lubricant supplied. A high flash-point steam oil was used, and provided the oil flow was sufficient to effect good internal sealing, the MVE efficiency remained comparatively high. The oil formed an emulsion with the exhaust condensate and so separation of the oil and water was virtually impossible. The emulsion was used as the lubricant with some success, but because it was not possible to design and build an effective separator, it was not known whether the emulsion would become stable or not. However the lubrication problems can be solved by appropriate technology, and a saturated-steam Rankine cycle using a MVE remains a very attractive proposition.

7.5 The Lobe MVE using Compressed Air

In view of its untried design and novel configuration the lobe MVE performed reasonably well. An oversight during the design stage gave rise to an excessive clearance volume (occupied by the springs) which meant that an upper limit of expander efficiency of ∿80% was imposed by the cut-off efficiency. Obviously $\eta_{cut\text{-}off}$ can be made greater by correct design in which case the expander efficiency would rise to ∿60% instead of the measured ∿48%. An independent test of the end seals indicated excessive leakage. This arose because the seal was of unbalanced design and not as closely fitting as required. An effective balanced seal should be able to reduce this leakage to near zero thus further improving efficiency. If γ_L were reduced to 50% of its average measured value, the expander efficiency would rise to ∿73%. Other minor design and/or material changes should also be able to enhance its efficiency.

One other aspect which affected its measured performance was that the freezing temperatures caused by the expanding air meant that the oil viscosity was higher than desired in spite of using a comparatively low viscosity oil.

With appropriate modifications the lobe MVE could realistically be expected to attain efficiencies in the 70-80% region.

7.5 The Rankine Cycle Optimisation

The optimisation examples given serve more to illustrate the fact that a definite optimum exists for an ORC using positive-expansion devices. With properly validated models, the optimisation could be extended to cover other variables, but those given are the major factors affecting the ORC efficiency.

The dramatic effect of leakage is well-illustrated by the second example, indicating that efforts made in the direction of minimising leakage (even at some cost to the mechanical efficiency) offer considerable scope for performance improvement.

8.0 Conclusions

Two types of MVE have been extensively investigated using both theoretical and practical approaches. A conventional MVE has been tested using R11 or steam as the working fluid and a lobe MVE has been tested using compressed air.

From the analyses, the performance of a MVE can be expressed in terms of various efficiency factors which combine to give a resultant expander efficiency. Of these factors the leakage coefficient and mechanical efficiency are dominant.

The ORC using a MVE (or any other positive-displacement expander) has been shown to have a definite optimum depending on the mechanical and leakage characteristics of the expander.

It may therefore be concluded that:-

- Adequate theoretical models have been devised to describe and predict the performance of both conventional and lobe MVE's.
- MVE performance is primarily affected by its design with respect to mechanical friction and leakage.
- An ORC may be optimised for maximum efficiency once the working fluid and expander characteristics are known.
- A conventional MVE appears capable of attaining 70→80% efficiency.
- A lobe MVE appears capable of 70→80% efficiency if effective positive internal sealing is employed.

The results indicate that the MVE, by appropriate design and development, can be a cheap, robust, quiet and efficient expansion device. It is therefore recommended that further design and development investigations of both conventional and lobe MVE be pursued. Results from this research could be expected to be near-term with subsequent application to waste-heat recovery systems and other low temperature bottoming cycles. It is further recommended that studies be made into larger capacity MVE's, i.e. in the 50→500 kW power range.

9.0 References

1. TEICHMANN, O.E.: "Analysis and Design of Air Motor"; Hydraulic and Pneumatic Power and Control Design, Performance Application, McGraw Hill Book Company, 1966

2. HUSSEIN, M.E.: "Low Grade Energy Engines"; Cranfield Institute of Tech., Ph.D Thesis, Sept. 1981 (Restricted)

3. CURRAN, H.M.: The Use of Organic Fluids for Rankine Engines" 15th IECEC, Seattle, Washington, pp. 985-991, Aug. 1980

DISCUSSION

Question - Mr Zegers, CEC : What are the problems with the feed pump in the ORC engine ?

Answer Pumping of the chemical fluids merits close scrutiny, since this operation is a major consumer of the expander power, ranging from one-fifth to three-eights of the total power output. At these levels of parasitic demand, direct drive from the expander shaft is preferable to reduce the losses. With the high ratios of pump work to expander work, inaccurate determinations of the property values in the supercritical ranges would be made. Subsequent incorrect estimates of pump work, could lead to totally false evaluations of the effectivenesses of alternative candidate fluids, of the net power output and consequently of the cycle's overall efficiency.

The required liquid feed pump incorporated in the organic Rankine cycle had to fulfil special demands dictated by the unique characteristics of the cycle's working fluids and components.

During the design of the feed pump to be used in the test rig at Cranfield, I was faced with the following example.
The cycle was designed to use R-11 as the working fluid, to work between 30° C to 150° C and to have a 5 kW output.
The demands to be met by the feed pump were :
- To have a high pressure differential of $\Delta P \simeq 20$ bar,
- To have low mass flow rate of $\dot{m} < 0.2$ kg sec^{-1},
- To excert a speed of ≃ 3000 rpm for direct coupling with the expander shaft,
- To ensure low noise level during operation to avoid cavitation of the pump.

Question - Mr Macchi, Politecnico di Milano/CNPM : I found the presentation of Dr Hussein very interesting. However, he stated that turbines must operate at fixed conditions, while our experience demonstrated that reaction turbines can have a satisfactory behaviour at part load. As an example, we have tested a small solar engine, and have found very high turbine efficiency in the range 1 - 4 kW, i.e. for a power range operation from 25 to 100 %.

Answer Thank you very much. I am very happy to get to know about the testing of your small turbine of 1 - 4 kW; My question however is what is the type of your turbine and what is the speed variations during that testing ?

Answer - Mr Macchi : I will send you this information.

Question - Mr Zegers, CEC : In heat pumps with sliding vane compressors, the lubrificating oil of the compressor is dissolved in the working fluid (10 - 15 %) and decreases the heat output of the heat pump up to 30 %. In your multivane compressor do you have the problem of lubrificating oil dissolving in the working fluid and how does this effect the operation of the ORC engine ?

Answer I have this problem since oil is completely miscible with R-11. Hence, it was very difficult to separate them by using the conventional oil separator, or even, a specially designed one. In my opinion, the problem of separation is worse after the expansion process rather than the compression process. The high outlet pressure which exists at the exhaust of the compressor increases the required driving force for the separation process. The wet vapour, at the exhaust of some expanders, handling especially what is called, a wet expanding working fluid, create a lot of problems for the separation (R-11 is a typical example). I have handled this problem by mixing the working fluid, R-11 - R-113, with 5 % (by weight) Clavus oil 68. No changes of the thermodynamic properties of R-11 and R-113 had been noted and the test rig had been in continuous operation for at least 1000 hours.

Question - Mr Hopmann, MBB : 1. There are typical advantages and disadvantages of the expander type prime movers and turbines with respect to power range, characteristics, partial load behaviour, etc. What is the typical advantage of the MVE compared to the screw expander, which can be manufactured from commercially available components and which has shown efficiencies higher than 85 % in operation of ORCS ?

2. Do you think that the MVE has commercial potential in the 5 - 50 kW power range in industry ? In solar energy the ORCS with flat plate collectors/ORC system may achieve system efficiencies not higher than 3 %. Photo-Voltaic has already demonstrated 10 % and higher. What is your statement with regard to cost effectiveness/economics ?

Answer 1. The typical advantages of the MVE compared to the screw expander are the following :

- better power to weight ratio (especially for level of power below 25 kW output),
- the MVE is an easy breathing device, which means it has a higher volumetric efficiency and lower losses,
- the MVE is simpler and cheaper in production,
- the MVE has flat characteristics for wider working conditions and lower speed,
- I believe that the introduction of an oil free machine for the MVE will be rather easier.

2. Yes. I think that the MVE has commercial potential in the 5 - 50 kW power range in industry. The ORCs in solar applications has the following advantages over the PV systems :

- The technology available in the developing countries could handle the ORC easier, especially for manufacturing purposes. It should also be stated that the manufacture of PV still goes through considerable problems, even in the industrialised countries.
- The ORC could produce AC power which could be used for several applications. The PV could only produce DC output which requires special devices so that it could be used further. It may require an alternator which would result in some losses and a cost increase.
- The efficiency of the ORC is about $\simeq$ 6 - 7 % but this could improve if the heat rejected from the condenser is used for different applications.
- The cost of the ORC is still lower than that of the PV, and it will be kept lower, for at least, another 10 years. However, the cost of ORCs could become even lower if they are produced in the developing countries.
- Further R & D on ORCs would help the promotion of different technologies in the developing countries. The introduction of new different techniques such as the introduction of the solar pond, and the recovery of waste heat using an ORC could be easier and more economic.

TABLE 1 Rankine Cycle Organic Working Fluids

* Mixture fractions on mole basis

FLUID	CHEMICAL FORMULA*	MOLECULAR WEIGHT	CONVENTIONAL DATA FOR CRITICAL PT. °C/kPa	FREEZING PT. °C	APPROX. THERMAL STABILITY LIMIT °C	VAPOUR PRESS AT 40°C/kPa
R-11	CCl_3F	137	198/4413	-111	120	159
R-21	$ChCl_2F$	103	178/5102	-135		276
R-22	$CHClF_2$	86	96/4978	-160	200	1455
R-113	$C_2Cl_3F_3$	187	214/3441	- 35	175	76
R-114	$C_2Cl_2F_4$	171	146/3261	- 94	175	317
R-133a	$C_2H_2ClF_3$	118	152/4068	-106	200	310
CP-9, (a)	$C_{15}H_{16}$	196	522/2448	- 55	370	
CP-25, (b)	C_7H_8	92	321/4254	- 95	480	7
CP-27, (c)	C_6H_5Cl	113	359/4523	- 55	320	7
CP-32, (d)	C_5H_5N	79	347/5633	- 42	370	7
CP-32/ Water, (e)	$0.23C_5H_5N$ $0.77H_2O$	33	366/8964	- 18	400	14
CP-34, (f)	C_4H_4S	84	307/5461	- 40	290	21
FC-75, (g)	$C_8F_{16}O$	416	227/1607	- 62	320	7
FC-88, (h)	C_5F_{12}	288	150/2131	-115	200	140
F-85, (i)	$0.85CF_3CH_2OH$ $0.15H_2O$	88	240/6412		290	21
Dowtherm A, (j)	$0.265(C_6H_5)_2$ $0.753(C_6H_5)_2O$	166	499/3241	- 48	370	
Biphenyl	$(C_6H_5)_2$	154	498/3289	- 69	370	
Allied P-1D	$C_{10}F_{22}O_2$	570	243/1186	- 85	370	2

(a) Monoisopropyl biphenyl $(C_6H_5)(C_6H_4)-CH(CH_3)_2$
(b) Toluene
(c) Monochlorobenzene
(d) Pyridine
(e) Azeotrope of pyridine and water
(f) Thiophene
(g) Perfluoro-2-butyltetrahydrofuran
(h) Perfluoropentene
(i) Fluorinol (Trifluoroethanol/ water mixture)
(j) Eutectic of biphenyl and phenyl ether

TABLE 2 Comparative Study of Suitable Working Fluids

	R-11	R-113	R-114	FC-75	FC-88
Cost and avilability	cheapest and available	relatively cheap ($\sim$twice cost of R-11) available	as for R-113	expensive and available	expensive and doubtful availability
Critical temperature (°C)	198	214	146	227	150
Heat transfer rate	average	average	low	high	near low
Pumping requirement	average	average	low	high	average
Rankine efficiency	high	high	average	low	low
Normal boiling point (°C), i.e. at atmos-pheric pressure	23.8	47.6	3.8	102.6	30.0
Vapour pressure (Nm^{-2}) at 40°C condenser temp.	1.59×10^5	0.76×10^5 (vacuum)	3.17×10^5	0.07×10^5 (vacuum)	1.4×10^5
Thermal stability	up to 120°C in the presence of oil	superior, up to 149°C	?	acceptable, up to 200°C	excellent, up to 176.7°C

A recent communication with the "3M" Company, UK branch, the manufacturer of FC-88, confirmed that this has been withdrawn from the market due to its high cost. Similar chemical compounds, which will be cheaper, will be available under the names of FC-72 and FC-87

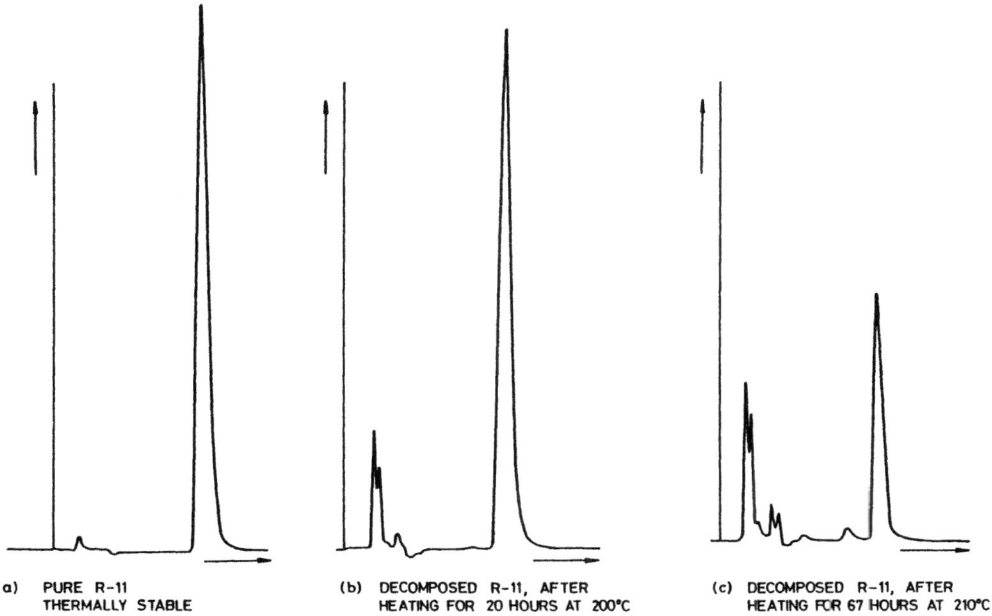

FIG. 1 THERMAL STABILITY TESTS (USING A GAS-CHROMATOGRAPH)

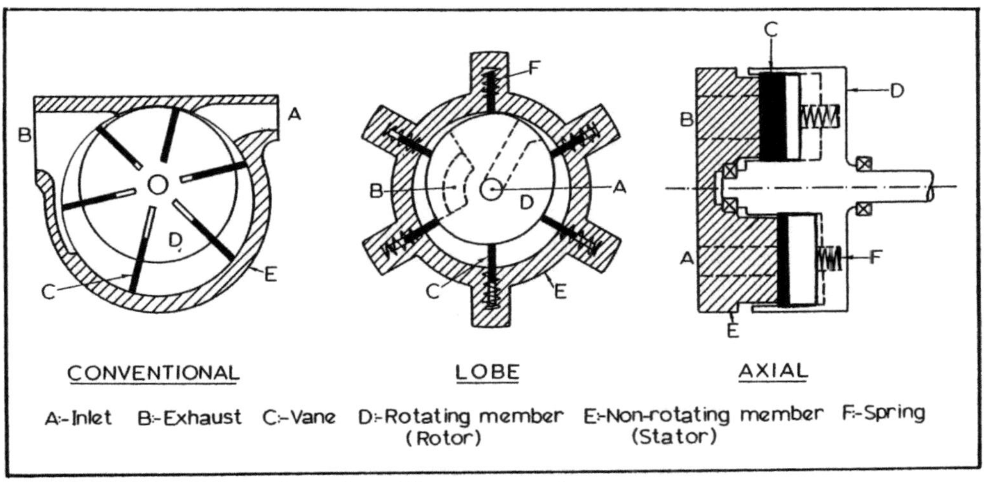

FIGURE 2 Three designs of multi-vane expander

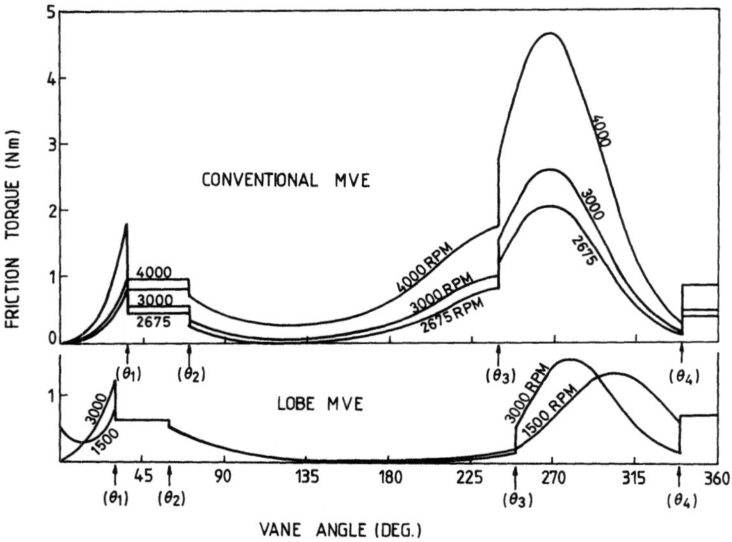

FIG. 3 COMPUTED INSTANTANEOUS VANE FRICTION TORQUE

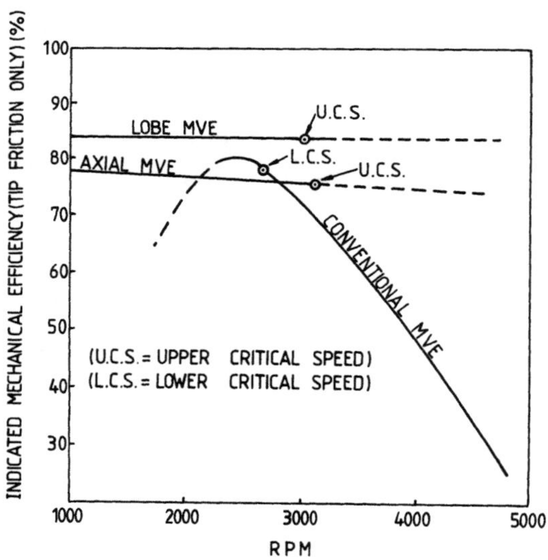

FIG. 4 COMPUTED MECHANICAL EFFICIENCY/SPEED FOR THE THREE MVE DESIGNS

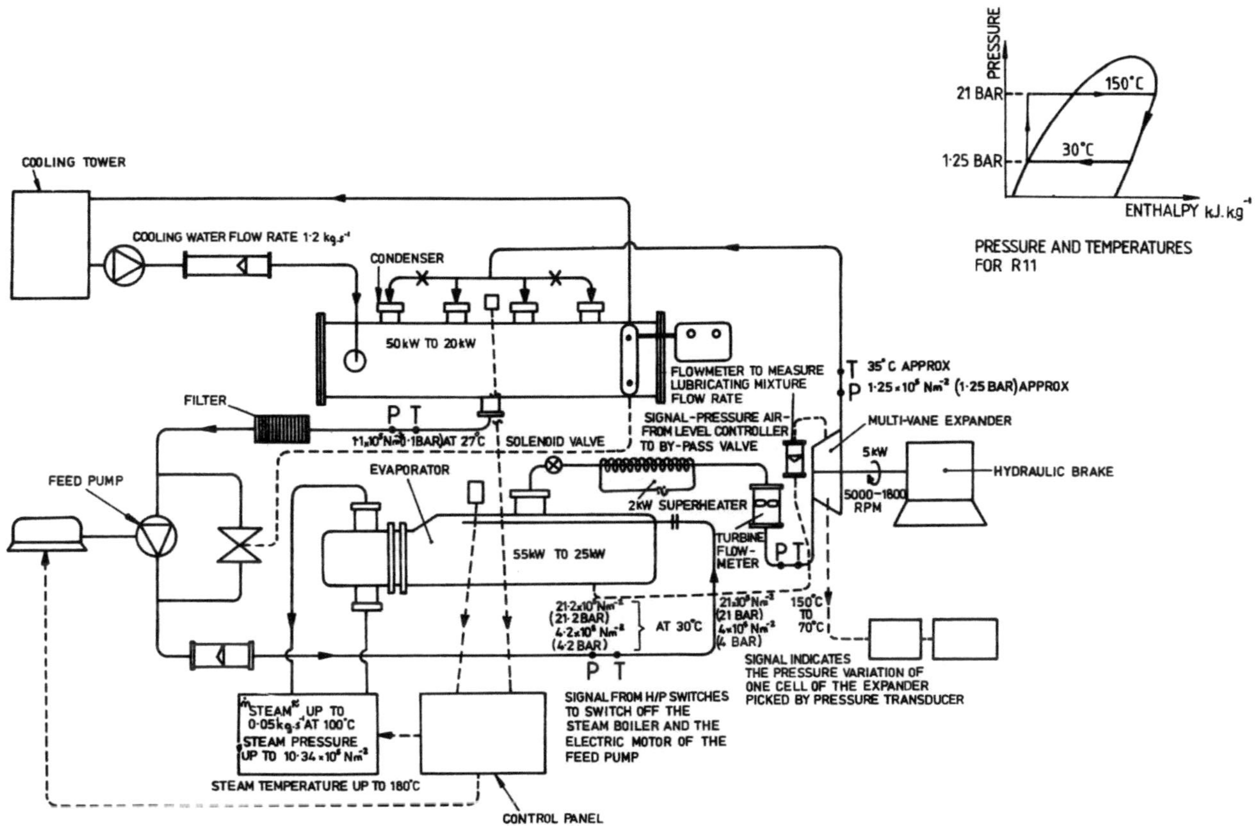

FIG. 5 REPRESENTATION OF THE RANKINE CYCLE TEST RIG

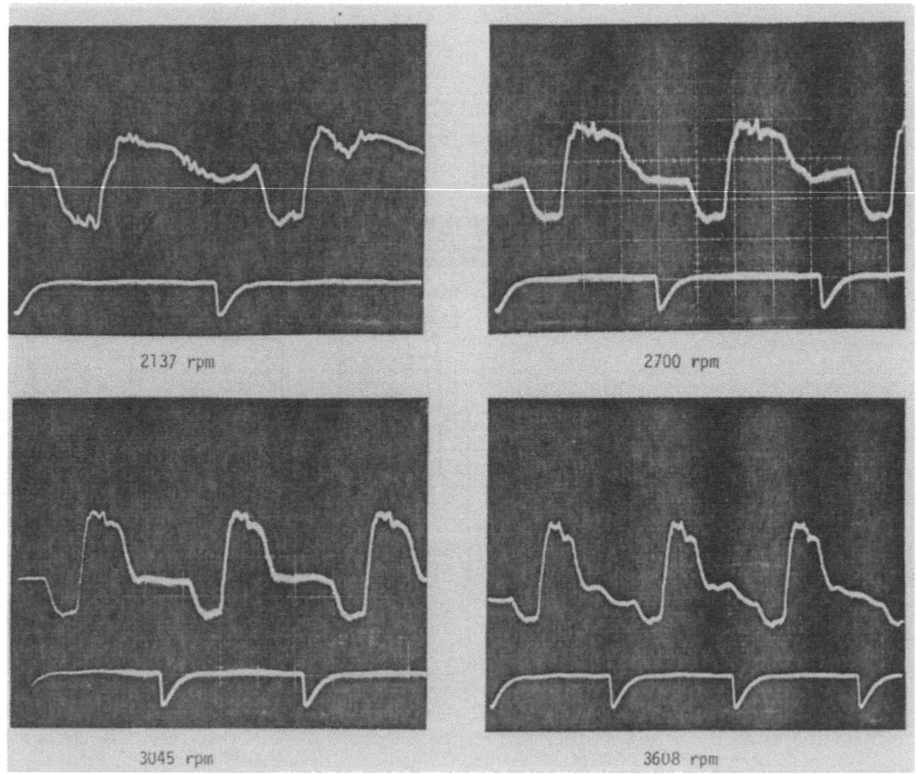

FIGURE 6 CELL PRESSURE VARIATIONS FOR DIFFERENT SPEEDS

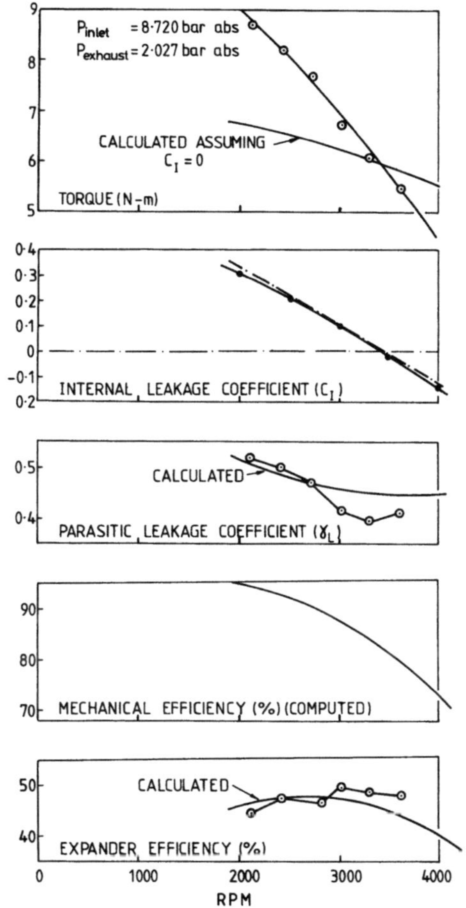

FIG. 7a PERFORMANCE OF THE CONVENTIONAL MVE USING R11

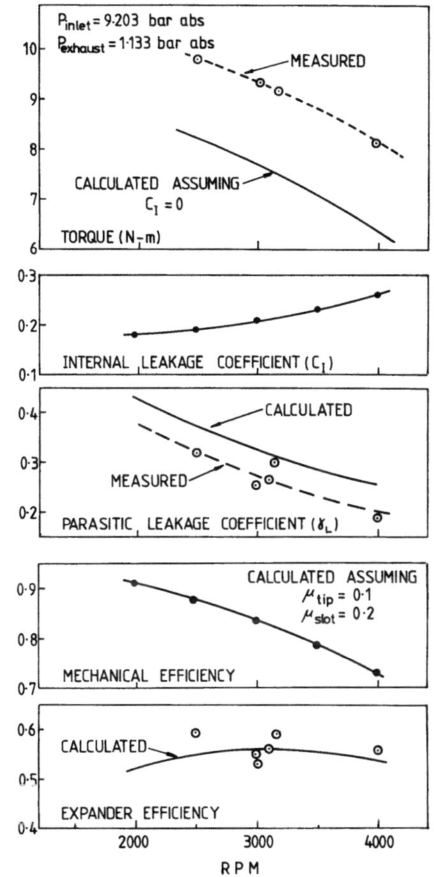

FIG.7b PERFORMANCE OF THE CONVENTIONAL MVE USING STEAM

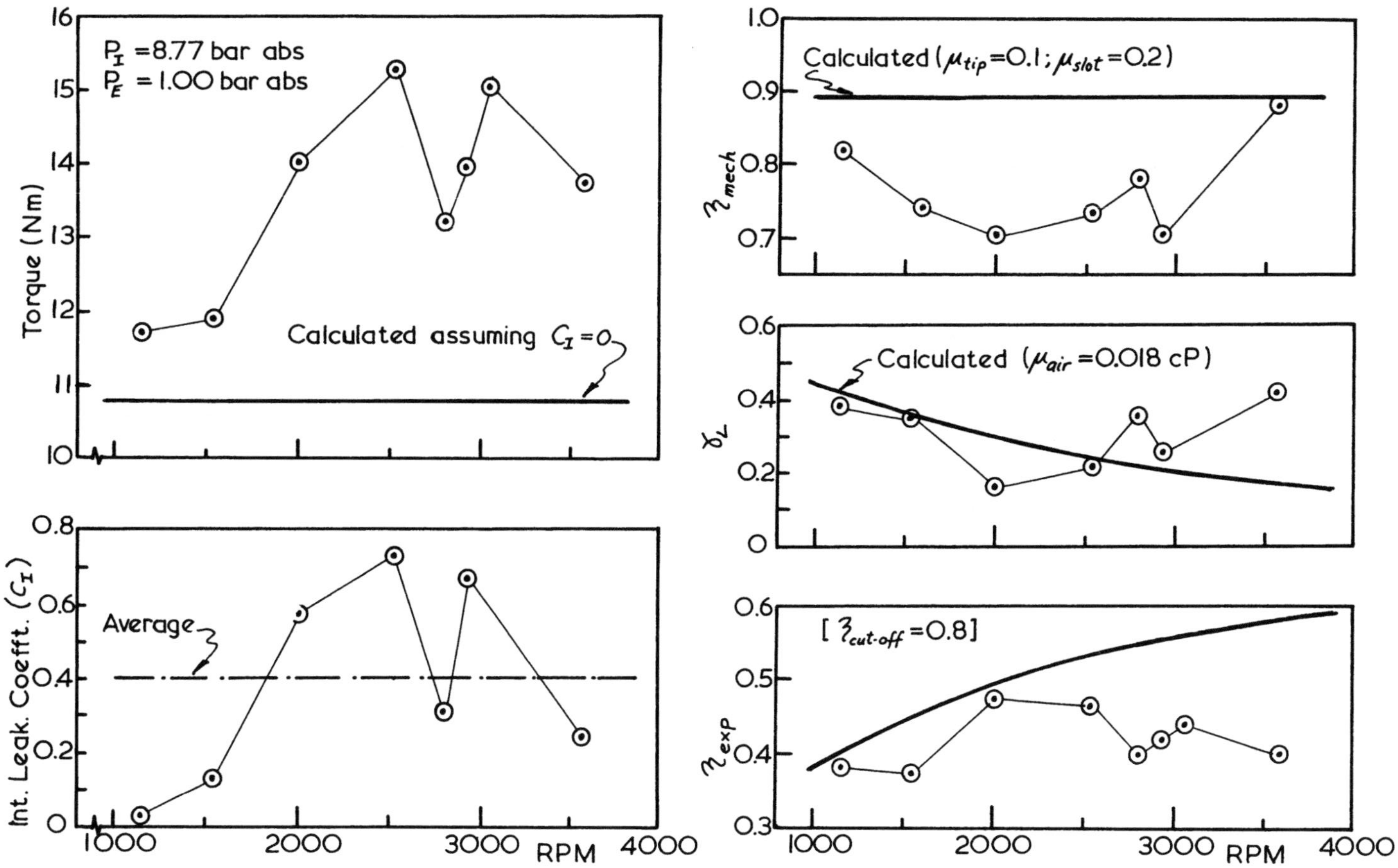

FIGURE 7.c. PERFORMANCE OF THE LOBE MVE USING COMPRESSED AIR

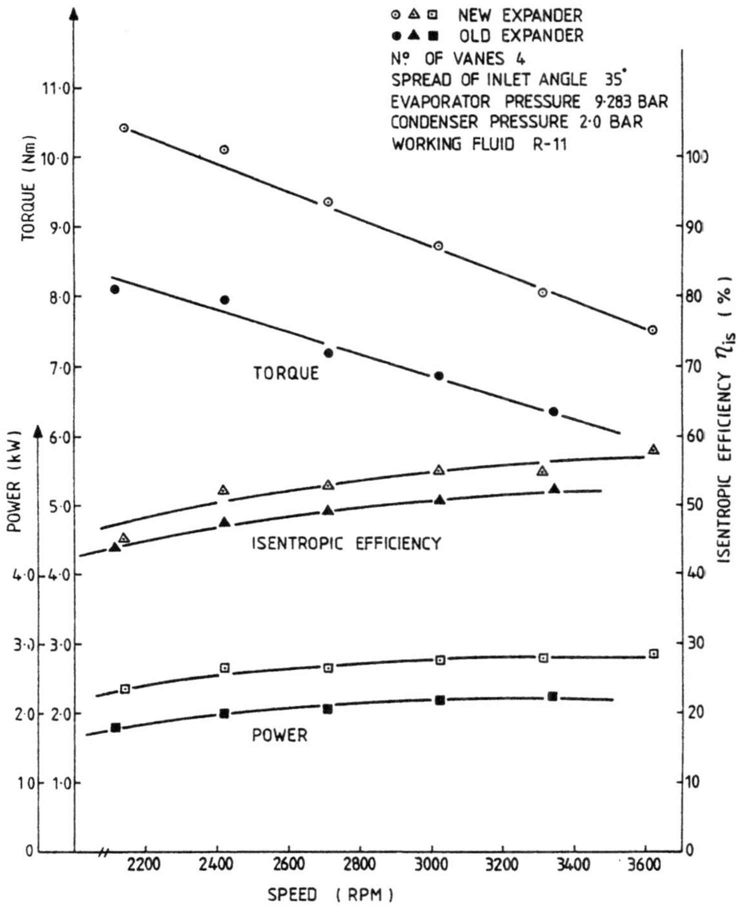

FIG. 7d MODIFIED MVE PERFORMANCE CHARACTERISTICS

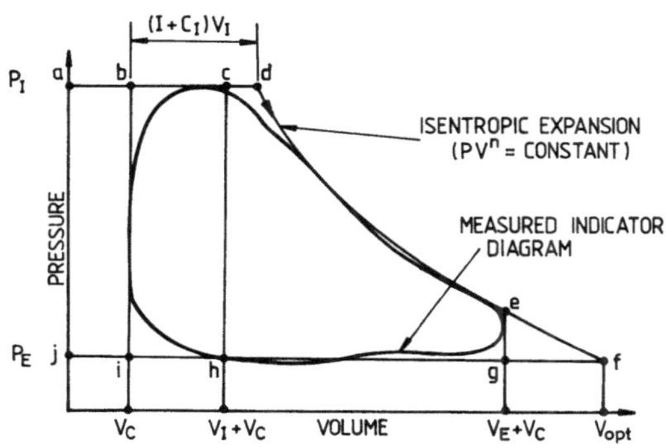

FIG. 8 MULTIVANE PRESSURE-VOLUME DIAGRAM

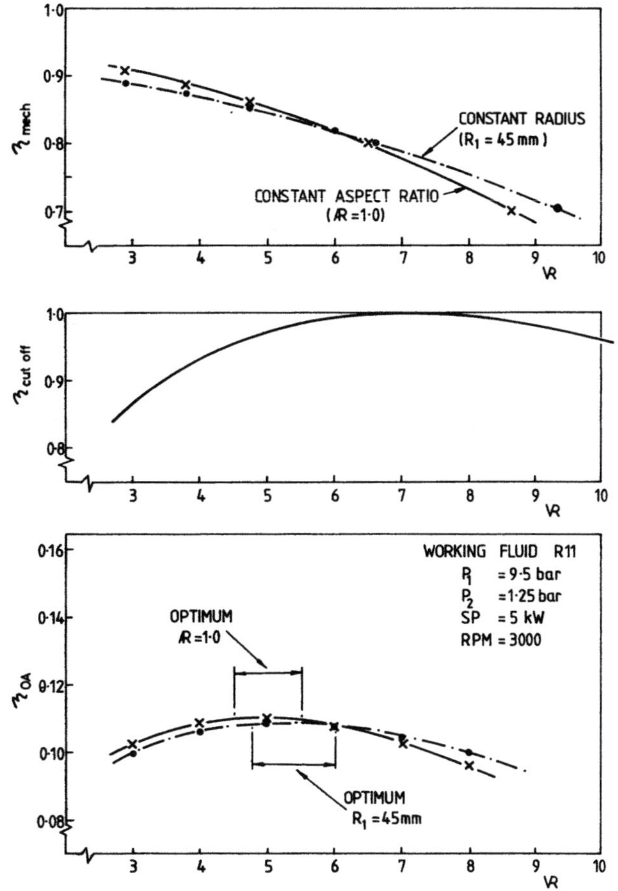

FIG. 9 CONVENTIONAL MVE - CONSTRAINED OPTIMISATION

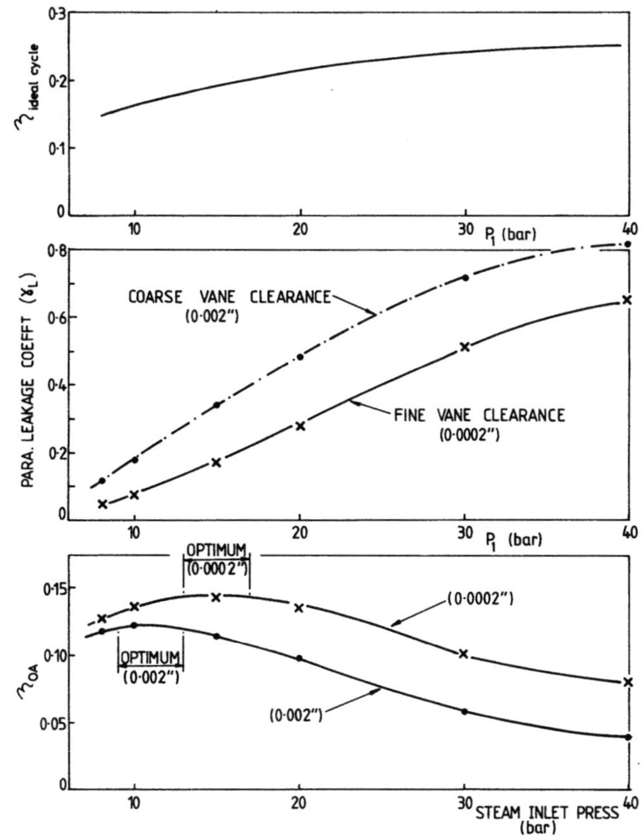

FIG. 10 LOBE MVE - UNCONSTRAINED OPTIMISATION (DRY SATURATED STEAM)

CONFIRMATION OF THE ADVANTAGES OF A THERMODYNAMIC CYCLE USING SULFUR WITH THE HELP OF AN EXPERIMENTAL FACILITY.

Authors : S. GOLDSTEIN, B. VRILLON.

Contract number : EE-D-3-341-F.

Duration : 36 months 1 July 1980 - 30 June 1983

Head of project : S. GOLDSTEIN, Commissariat à l'Energie Atomique

Contractor : Commissariat à l'Energie Atomique

Address : CEA - DEMT - SMTS - TTMF.
91191 GIF-SUR-YVETTE - FRANCE.

Summary

The problem of the corrosion of steel at high temperatures by sulfur has been solved on small samples in laboratory. This fact opens the way to the use of sulfur as a thermodynamic fluid for energy production between 750°C and 500°C. Based on spectral data, the overall efficiency of a binary cycle sulfur-water plant is attractive (∿ 60%). The present contract concerns the experimental confirmation of the interest of such a cycle. In this report we summarize the work done in the field of thermodynamics and give the present status of progress of the experimental work.

I. Introduction.

The major difficulty which appears if one wants to use the sulfur as a thermodynamic fluid is the corrosion of the structural material by the sulfur. A good material must :

- Resist to corrosion for several years
- Have good mechanical properties up to 750°C
- Be cheap.

Such a material has been found by the "Laboratoire de l'Ecole des Mines" at Corbeil under a contract of "Electricité de France". The present contract is not concerned by this study and the description of this research will not be made here. We just keep in mind that this material is steel, coated with a cement made of aluminium and titanium. This coating has to be oxidized prior to its use to generate a film of alumina which stops the penetration of sulfur. Besides a low concentration of oxygen has to be maintained in the sulfur to regenerate the alumina film. More details on this study can be found in the decision report issued on July 1st 1982.

These corrosion studies performed on small samples in laboratory has opened the way to the use of sulfur as a thermodynamic fluid, but the technological problems set by the industrialization of the process are far from being solved. This appeared clearly when we started to design the experimental loop described below, and is the reason why it seemed to us useful to perform preliminary tests on small scale experiments, before to start the operation of the loop.

For such a subject where only few work has been carried on till now, there are many topics to develop and it is essential to fix methodic objectives and define a programme of work. As a first step, we decided to perform some physical experiments on the sulfur in liquid and vapour phase. The objective of these experiments is the collection of physical data without trying to solve any industrial problem. For this reason we choosed for the conception of the loop, solutions which are likely to succeed even if they are far from an industrial design. Later on the loop will be useful to perform technological tests with industrial components and to help the design of a pilot plant.

The loop named ESTHER is made of two parts. The low temperature part (at 140° or 300°C) which contains the tanks, valves and pump and is the alimentary part of the loop ; the high temperature part (temperature comprised between 140° or 300°C and 750°C) which contains only heating and cooling elements without any machinery.

The plan of the report is the following.

§II. Conclusions of the preliminary thermodynamic study
§III. Preliminary tests
§IV. The low temperature section of ESTHER
§V. The high temperature section
§VI. Experimental programme
§VII. Related research programmes
VIII. Conclusion.

II. Conclusions of the preliminary thermodynamic study.

The thermodynamic study is made in ref[1]. This report contains mainly two parts :

* A review of the physical properties of the sulfur. This review puts in evidence the lack of experimental data on the liquid phase above 350°C and the fact that the physical properties of the vapour are deduced from

spectral data of the molecules (the vapour contains seven different molecules : S2 to S8).

* The calculation of species concentration in the vapour and build up of the thermodynamic diagramm. This diagramm is shown fig. 1. Simple calculations of the efficiency ε of a combined sulfur water binary cycle, confirm the advantage of this cycle ($\varepsilon \sim 60\%$). It is interesting to note that.

- The expansion of the saturated vapour leads to superheated vapour.
- The isenthalpic curves of the vapour are not horizontal as for a perfect gas, but has an important slope. This slope is due to the chemical reactions which occur between the species. This can also be seen on the constant pressure heat capacity curves [1].

Internal and external irreversibilities can occur during the adiabatic expansion of the vapour. The external irreversibility essentially due to friction has not been considered here. An internal irreversibility can occur if the chemical reactions are performed with a finite velocity, i, e if the chemical species are not in equilibrium with temperature and pressure at all the stages of the expansion. To evaluate this effect computations has been done for expansions in equilibrium or with fixed chemical species. The result is shown fig 2.

. AB represents an expansion in equilibrium.

. AB' an expansion with chemical species kept constant (those calculated at A).

. If we suppose that the condensor works at a given temperature (i-e 480°C), several cases have to be considered :

-a- The expansion is in equilibrium ; then the cycle swept is ABD and there is no loss in the efficiency.

-b- The expansion is not in equilibrium, but nevertheless the chemical reactions occur in the condensor. The cycle swept is AB'CD and the loss in the efficiency is minor (less than 1%).

-c- The expansion is not in equilibrium, and the chemical reactions do not occur in the condenser. They occur somewhere else in the cycle. The cycle swept is then ABD, but with a final pressure of 0.5 bar in the condenser. The loss of efficiency is comprised between 4% and 5%. In a sense, the sulfur behaves like a chemical accumulator which releases during the expansion the stored chemical energy. The recommendations of the thermodynamic study concerning the physical measurements are :

- To measure the heat capacity, conductivity and viscosity for the liquid phase between 350 and 750°C. Among them the viscosity is the more important because it has a wide range of variation and its behaviour cannot be predicted.
- To check the diagramm in equilibrium.
- To measure the time constant of the chemical reactions.

III. Preliminary tests.

Handling the sulfur at high temperature and pressure is not a routine duty. Even if a design of the ESTHER loop that should work is found, it is wise to test before some of its components and instrumentation on a separate and smaller scale experiment. This experiment is shown fig 3. It comprises an owen whose temperature can be raised up to 750°C, which contains the component to test. This component is fed by a melting pot through valve V4. The preliminary tests will permit to check the following points :

. tightness of valve V4.

. well standing up of the pressure sensor P

Resistance to corrosion of the tube, flange sealing joint and sulfur thermocouples in the owen.

Previous oxidation will be done and a partial pressure of SO_2 will be kept in the section during the experiments.

Besides many sensors are possible to use to measure the level in the tanks of the loop (capacitive, ultrasonic, pressure ...). To chose the cheapest and more reliable one they will be tested and calibrated on a small tank containing liquid sulfur at 140°C or 300°C.

Both these experiments are designed and ordered and should start in October.

IV. The low temperature section.

A sketch of the low temperature section of the loop is found fig. 4. It comprises seven tanks, fifteen sulfur valves and one volumetric pump. Briefly, this part of the loop works as follows :

. The sulfur is melt is the tank R_1 and flows through V_0 and V_1 in R_2.

. When R_2 is full, the pump starts operating and forces the sulfur through the hot part of the loop which is comprised between V_6 and V_7 and not sketched on the figure.

. While R_2 is emptied, R_3 is filled by the sulfur which turns back through V_7 and V_8.

. When R_3 is full and R_2 is empty V_2 is closed, V_3 is opened and a new cycle begins.R_2 and R_3 are filled and emptied by turns.

. R_4 and R_5 are smaller tanks which serve as air buffer to avoid pressure oscillations during the pumping.

. R_6 is a discharge tank for the liquid sulfur.

. R_7 is a discharge tank for the vapour and gases. Il also serves to create a vacuum in R_2 or R_3 when the sulfur vapour expands to a final pressure below the atmospheric pressure.

All this part of the loop is designed to work at a temperature of 300°C. As it will be said in the next paragraph. If there are no major problems to get through the peak of viscosity in the elements of the hot part of the loop, it will be possible to work at 140°C.

An argon circuit connected to the tanks R_2 to R_5 serves to adjust the pressure in these tanks.

The whole loop is ventilated and the discharge gazes purified before to be released in the air.

This design of the loop allows several working conditions i, e :

. with the pump - liquid or two phase flow - expansion to a pressure above 1 atm.

. with the pump - two phase flow - expansion to a pressure below 1 atm.

. without pump - expansion to any pressure.

The case without pump has been considered because there is no guarantee that the pump works at 300°C.

By now the main components of the loop have been ordered and the detailed study of the design is in progress.

V. The high temperature section.

The simplest conception for this section would be a tube heated and cooled from the outside as it is described in [2]. By now the internal protective coating of alumina is feasible on short tubes with a maximum length of one meter. This has been actually done on a 10 mm diameter tube by "Société HEURCHROME" and the coating had a normal visual aspect.

If this section of the loop has to be assembled with such short tubes we need to find a tight and corrosion resistant joint to insert between the flanges. We have not yet found such a joint. As a matter of fact, a steel joint protected by the alumina coating would not be convenient because the coating does not resist to crushing nor a sheet of gold would do because gold is melt in the liquid sulfur above 600°C. One purpose of the preliminary tests is to find out such a sealing joint. Welding is an other option for assembling the tubes, it has to be checked that it does not destroy the internal coating. For these reasons we preferred to choose for the high temperature section of the loop a solution with internal insulation and cold steel tubes (∿ 300°C). The loop is made of elements assembled in series. A sketch of one element is shown fig 5. It is made of a steel tube insulated internally with alumina and cooled externally. The sulfur flows in the annular space between the alumina and an inner quartz tube which contains an electrical heating. The quartz tube is stuck on the steel tube at one end which is maintained cold with the help of an air cooling circuit. Due to the bad conductivity of the sulfur the thickness of the annular flow is 1.5 mm. The same elements are used for heating and cooling. The electrical heaters are replaced by a pipe blowing air inside the quartz tube.

The test section will take place between the heating and cooling elements.

If another solution is not found after the preliminary tests (§ III), the same elements will be used for the two phase flow but with a larger annular flow thickness. The problem will be to avoid the condensation of the vapour due to the external cooling. The thermal study of the heating and cooling elements for liquid or two phase flow is in progress.

VI. Experimental programme.

Three groups of tests will be performed on the loop :

1. Running tests.

The high temperature part of the loop being by passed, check that all the components of the loop (valves, pump, regulation ...) work well at 140°C.

Insert between V_6 and V_7 one heating and one cooling element without internal insulation. Check the possibility to get through the peak of viscosity in these elements. The low temperature part of the loop works at 140°C, the heating element brings the temperature up to 300° and the cooling element back to 140°.

If we succeed with these running tests, the cold part of the loop will be operating at 140° for the next tests.

2. Liquid phase tests.

Before to study the expansion of vapour, the loop will run with liquid sulfur. Two types of measurements are possible.

. Heat capacity, with the heating elements.

. Viscosity by measuring a pressure drop in the test section. The thermal calculation and design of the test section are in progress.

3. Vapour expansion tests.

The measurement of pressure temperature and species concentration during an isentropic expansion presents difficulties. Construction difficulties because one would have to assemble a transparent quartz nozzle with a steel tube at 750°, and physical interpretation of the results because the expansion would not have been adiabatic and that there are no means to measure the temperature and pressure. The only possible thing to do is to measure the pressure and temperature drop at the ends of an isenthalpic expansion. This expansion can be obtained either with a nozzle (ANM on fig 1) or with a porous media (AM on fig 1). The expansion with a nozzle is more realistic because AN is the expansion that occurs in a turbine. The measurement of the final temperature gives information on the occurence of the chemical reactions during the expansion. For the case described above, for example, the final temperature is 750°C without chemical reactions and 560°C with chemical reactions. This temperature difference is measurable even if the expansion is not perfecty adiabatic.

VII. Related research programmes.

As it was mentioned in the introduction, there are many topics to develop about the use of sulfur as a thermodynamic fluid. Two research programmes are by now in progress.

1. Study of the solubility of SO_2 in the sulfur. This study is performed at the "Laboratoire d'Energêtique de l'Ecole des Mines de Paris" and is financed by the present contract. In the ESTHER loop or in an industrial plant, SO_2 will be introduced in the liquid phase and removed in the condenser. This study will be useful for the control of SO_2 concentration. It has started in september 1982.

2. Measurement of the species concentrations in the superheated vapour in equilibrium by Raman spectroscopy. This work is performed at the Centre National de la Recherche Scientifique of Thiais and is not financed by the present contract. The results of this research should confirm the calculations of species concentrations performed in [1] and consequently the thermodynamic diagramme. This work is in progress.

VIII. Conclusion.

This report summarizes the work done under the present contract. It gives the conclusion of the thermodynamic study and in particular the effect of the chemical reactions on the efficiency of the cycle. It emphasizes the fact that there are technological difficulties in the construction of the loop and that preliminary tests are necessary. Technological solutions has been found for the low temperature part of the loop. A solution with internal insulation is proposed for the high temperature part of the loop and is considered as the reference solution. It will be realized if no simplest solution is found at the end of the preliminary tests.

REFERENCES.

[1] S. GOLDSTEIN, N. JUIGNET, P. GREVERATH.
Elements de thermodynamique du soufre. Internal report DEMT.SMTS.TTMF. 82.24.

[2] S. GOLDSTEIN.
La boucle à soufre. Première étape le circuit liquide.
Internal report DEMT.SMTS.TTMF.82.01.

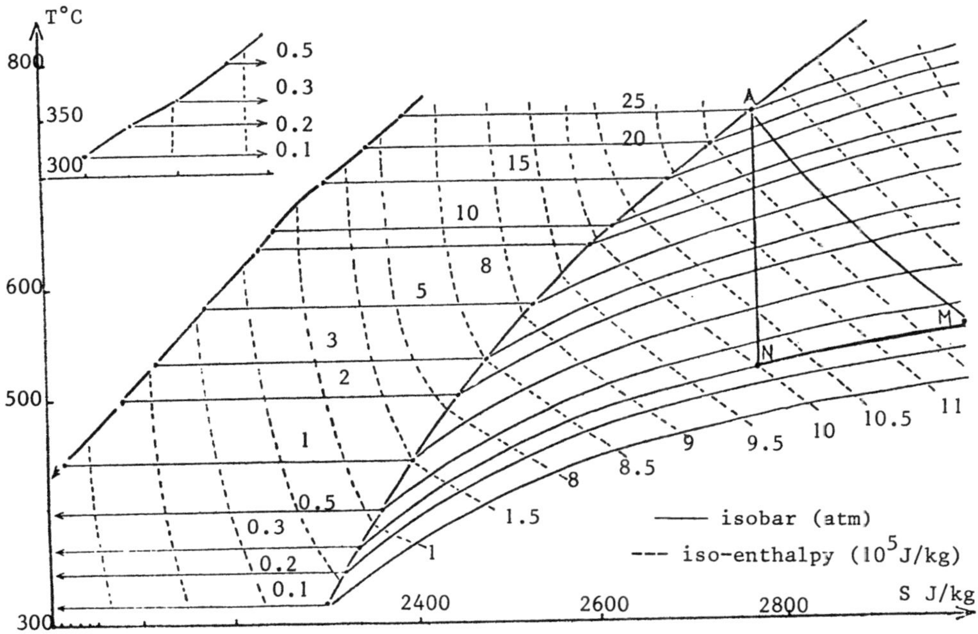

FIGURE 1 - Entropic diagram of sulfur.

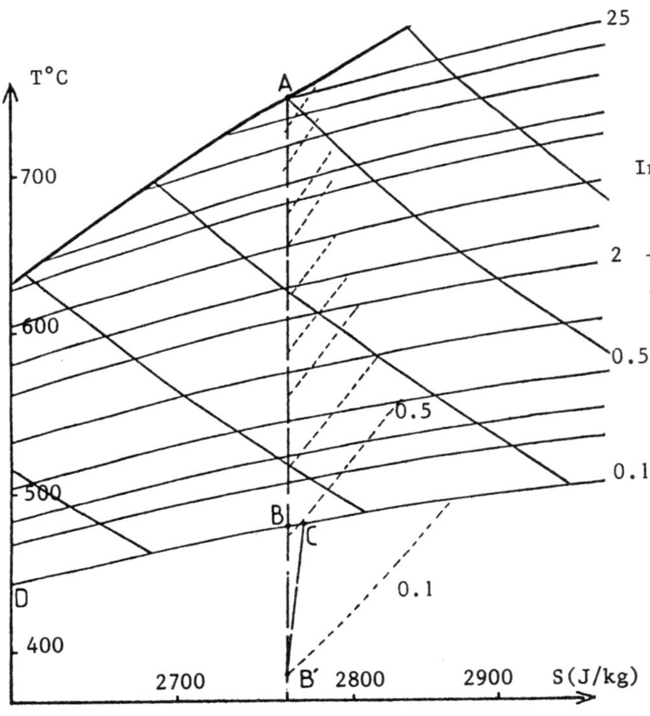

FIGURE 2 - Internal irreversibility.

DISCUSSION

Question - Mr Macchi, Politecnico di Milano : A sulfur cycle would operate at temperatures similar to the ones of a metallic liquid (Na, K, Hg) cycle. Which are the advantages of sulfur in comparison with the metals for this application ?

Answer Hg is expensive and toxic.
K-H_2O has the same efficiency as S-H_2O but with a maximum temperature of 850° C instead of 750° C and with lower pressure. The chemical reaction K-H_2O is explosive.
Sulfur is cheap, non toxic, has no explosive reaction with water and its T-S diagram combines well with a superheated water cycle.

Question - Mr Zegers, CEC : The use of sulfur as a working fluid in heat pipe heat exchangers between 300-500° C is, up till now, impossible due to corrosion. Would it be possible to apply your method of coating with aluminium and titanium to heat pipes so that the metal is not attacked by sulfur ?

Answer The coating method is based on a specific thermochemistry process applied to a steel support. It can protect against sulfur corrosion at temperatures as high as 750° C if a (very) small partial pressure of O_2 in sulfur is maintained in order to form a stable alumina layer.
In the case of the heat pipe this method would probably be effective. To confirm this applicability it would be useful to define precisely the general conditions (size, ...).

Comment - Dr Hussein, Cranfield Institute : The coating of heat-pipes could create problems specially due to the existance of foreign gases which tend to block the heat pipe and stop its performance (evaporation - condensation), I think the surface (inner) of the heat pipe must be smooth and clean.

Question - Mr Hupmann, MBB : Using sulfur as working medium for topping cycles do you really expect cost benefits by the higher cycle efficiency compared to the Brayton cycle ?

Answer The sulfur cycle is to be used with an external heat source (e.g. PFBC : pressurized fluidized bed coal furnace). It is not comparable to a gas cycle with internal heat generation, because the two cycles are not used in the same conditions and for the same purposes.
If for the two systems the heat source is external then the efficiency of the gas cycle is much lower than that of the sulfur cycle.

Question - Mr Bell, Cranfield Institute : What are the problems in starting and shutting down a topping cycle using sulfur ?

Answer These problems have not yet been considered. There will probably be problems related to the viscosity and solidification.

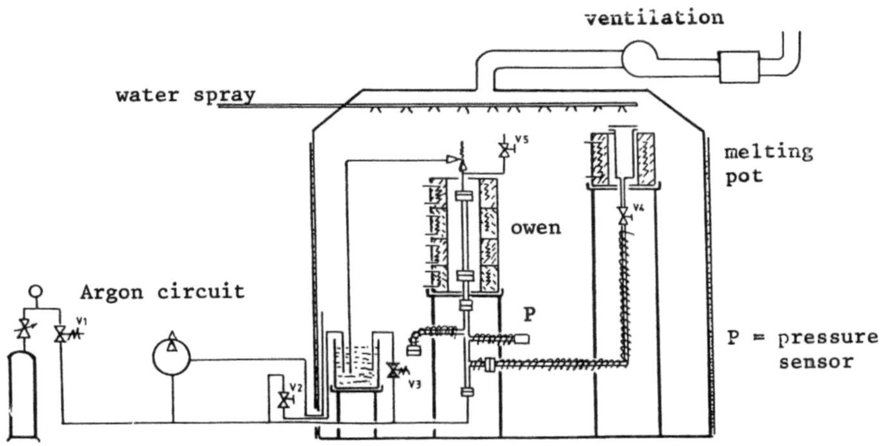

FIGURE 3 - Experimental apparatus for the preliminary tests

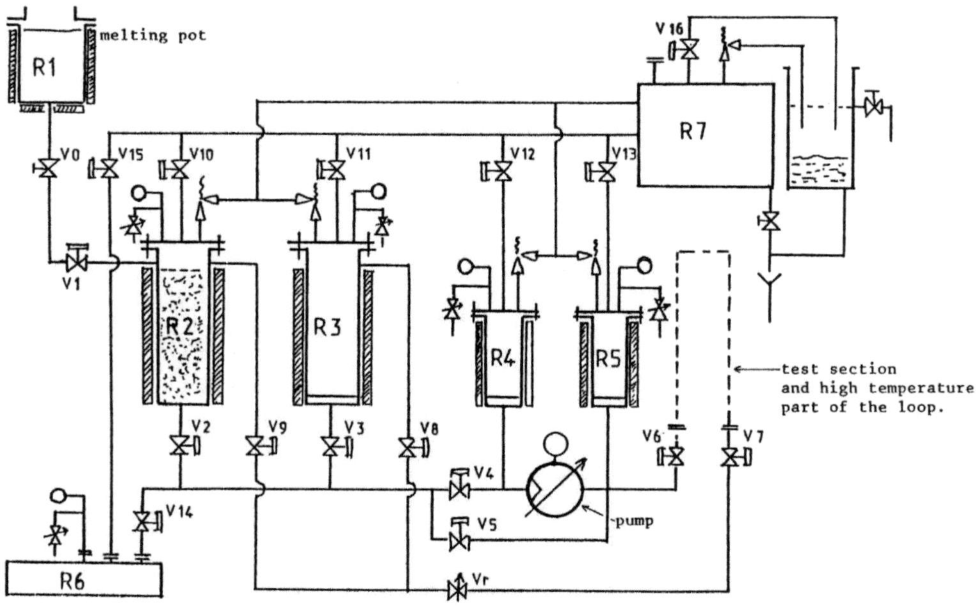

FIGURE 4 - Low temperature part of the ESTHER loop

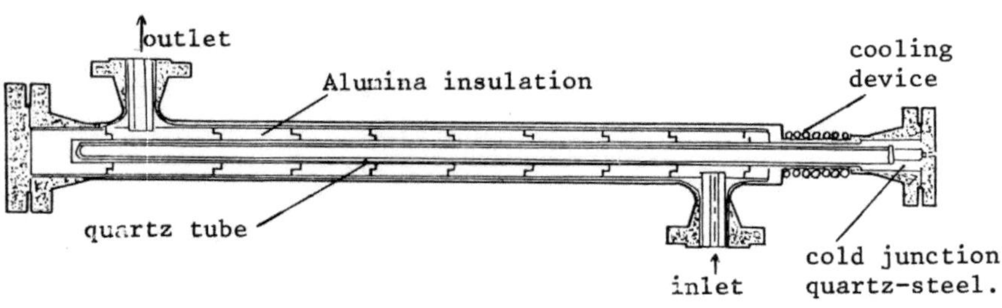

FIGURE 5 - Heating element

Energy Cascading Combined with Thermal Energy Storage in Industry

Authors: R.J. Wood, D.T.Baldwin, P.W. O'Callaghan and S.D. Probert

Contract number: EE-B-1-145-UK (N)

Duration: 24 months 1 November 1980 - 31 October 1982

Head of project: Dr P.W. O'Callaghan, Reader in Energy Management

Contractor: Cranfield Institute of Technology

Address: Cranfield, Bedfordshire, MK43 OAL, UK

Summary

The opportunities for energy conservation through the application of storage cascades has not previously been examined in sufficient detail to enable the overall technical and economic potential to be estimated. The objective of the research reported here, is the evaluation of this potential for UK industry.

A detailed survey of a representative sample frame of UK industry has shown that the technical and economic potential represents about 0.8% of the total energy use in the manufacturing sector of the UK economy.

Using technical constraints based on achievable power densities for a variety of storage technologies and economic constraints based on annual charge cycle frequencies, the feasible percentage of processes has been established for each storage technology. These results have been scaled using the 1980 reference energy use as a function of process operating temperature. Final scaling of the results to determine the UK industrial potential for storage cascades was obtained by evaluating the probability of two or more 'on-site' process cascade matches. The highest probability of a satisfactory cascade match occurs in the temperature range $100^{\circ}C \leqslant T \leqslant 250^{\circ}C$.

Steam accumulators appear to be able to contribute most to industrial energy conservation in intermittant processes. New generations of storage technology are far less significant. No processes were feasible for thermo-chemical storage systems due to the imposed power density constraint.

1.0 INTRODUCTION

In recent years thermal energy storage has been recognised as a potentially significant means by which the primary energy consumption can be reduced in domestic commercial and industrial space heating, in power generation and in industrial processes. On closer inspection of the available technologies however, it is clear that thermal energy stores do not save energy in their own right! In fact, energy and thermodynamic availability are always lost during a storage process. Resolution of this apparent contradiction can be understood because energy storage permits energy savings to be made by acting as an aid to the energy management of complex thermal systems.

In this respect the various methods of thermal energy storage form a potentially important set of "enabling" technologies, which could increase both the cost effectiveness and energy conservation potential of existing capital equipment.

The general aim of thermal energy storage is the improvement of the average to maximum power rating of "base-loaded" or "worst case" designed systems, with a consequent reduction of peak energy dissipating capacity and an improvement in the primary energy conversion efficiency of the system. When these aims are applied to industrial processes it becomes clear that thermal energy stores become attractive if they permit hitherto waste-heat streams to be considered as useful heat sources. In this context one must examine storage in a three dimensional space consisting of time, temperature and location. In addition storage components must also be considered as elements in complex thermal energy cascades in order to maximise the conversion efficiency of the primary energy input to a series of processes. The apparent complexity of both the storage and the coupled thermal systems makes decision taking concerning industrial potential and the technological developments required, necessarily difficult. Nevertheless, such information is desirable before meaningful judgments can be made about the development of many of the storage technologies currently under investigation. In addition, detailed information about industrial process characteristics should shed light on the most important criteria for the successful design of industrial thermal energy stores.

Today waste heat recovery via continuous "in stream" devices represent one of the highest priority energy conservation measures for industrial processes.

Current industrial applications of storage include high temperature regenerators in the metallurgical and glass industries, steam accumulators in the paper and process industries, and hot water storage in the low temperature process industries.

The objective of the research programme reported here, is to identify the technical and economic potential for energy conservation within UK industry for thermal cascades combined with energy storage.

The successful implementation of the appropriate technology should lead to reduced new plant requirement (or increased existing plant capacity), greater flexibility to achieve waste heat recovery, and short plant payback periods due to the potential for a large number of charge cycles per year.

1.1 Scope of the Investigation

The present study is separated into four phases:-

Phase 1

A comprehensive survey of UK industry in order to determine process information directly relevant to energy storage and the assessment of the current state of the art of industrial energy storage technology.

Phase 2

Analysis of the survey results in order to identify the general energy consumption characteristics of a wide range of industrial processes.

Phase 3

Analysis of the survey results in order to set technological and economic constraints and targets for thermal storage and cascade devices.

Phase 4

Refined economic analyses of industrial energy storage technologies including the energy saving potential, safety and environmental considerations and barriers to the successful implementation of such technologies.

2.0 Questionnaire Surveys

The first phase of the research programme was aimed at determining the energy use patterns, particularly relevant to thermal storage and cascading, within industry. A pilot questionnaire was initially circulated to 502 UK companies highlighting 147 thermally activated processes. A second and more detailed questionnaire, was subsequently circulated to 2746 companies within a sample frame constructed from the 'Classified List of Manufacturing Businesses' and segmented under the 'Minimum List Heading' system. A reference energy consumption for the sample frame was determined from the 1968 Census of Production Ref. (1) and later refined using the Digest of Energy Statistics 1981 Ref. (2) for the individual sector groupings between 1968 and 1980. The 1968 reference energy use for the sample frame was 1.08×10^6TJ (=36.6 MTCE). The 1980 reference energy use for the sample frame was 1.06×10^6TJ (=35.92 MTCE) which showed an overall decrease of 2% although major structural changes (up to 40%) had occurred between the sector groupings. The overall response to the survey was 7.6%, with the useful response representing 5%, of those questionnaires issued.

3.0 Results and Analysis

A structured data base was constructed from the completed questionnaires. This data base was subsequently analysed using the S.P.S.S. Ref. (3) computer program. This is designed to facilitate detailed analyses of questionnaire replies using statistical techniques: it also allows use to be made of multiple response replies.

3.1 Results and Analysis

The results for phase 2 of the programme are shown in figures (1-8) in the form of histograms for the main response categories of the survey.

3.2 Discussion of Phase 1 Results

3.2.1 Process Fuel Distribution

The process <u>fuel</u> used, figure (1), closely reflects current industrial practice in the UK. The category "sulphur" refers to one particular case in the chemical industry where sulphur is burnt to produce sulphur dioxide and the heat liberated is used in waste-heat boilers.

3.2.2. Process Heating Medium Distribution

The process <u>heating</u> media, figure (2), are dominated by combustion products and steam, which are typically rapid heat-transfer mediums and indicative of energy intensive processes.

3.2.3 Process Cooling Medium Distribution

The process <u>cooling</u> media, figure (3), are dominated by air (natural convection) and water. The special category "process" refers to those instances where a steam raising boiler is considered to be a primary process and is dissipating heat to a secondary process or series of processes.

3.2.4 Process Operating Temperature Distribution

The process operating temperature range, figure (4), indicates two major regions of interest. The first covers the range 50°C to 250°C, where the temperature requirements of processes increase in the ratio of approximately 1:5:3.5. The second block is applicable to high temperature processes in the range 251°C to greater than 1000°C, where the process temperature requirement increase in the ratio 1:1.5:3. Low temperature thermal energy storage systems are usually referred to as operating below 120°C. It is clear from the above results that a temperature of 250°C represents a natural boundary between high and low temperature regimes.

3.2.5 Process Cooling Temperature Distribution

The process cooling temperature range, figure (5), of most significance lies between 50°C and 85°C. There is some imbalance in the range <50°C because this has been used as a default group for those respondents who entered, for example, flue gas exit temperatures, which does not imply using a cooling medium in the conventional sense.

3.2.6 Process Operating Periods Per Day

Transient thermal processes can be characterised simply in terms of the number of times a particular process is taken from initial to final conditions in a fixed period of time. The most rational time period is represented by the working day, however, a working day can also be characterised by 8, 16 or 24 hour operation denoted by 1, 2 or 3 "shift" working, as well as smaller sub-divisions.

Figure (6) illustrates the results obtained for process operating periods per day (i.e. for the respondent's operating hours). It is clear that "continuous" processes dominate the distribution. However, it is accepted that some confusion over the term "continuous" did exist with the respondees. It is probable that many potentially useful sub-periods of process operation will exist within the large group of continuous processes.

3.2.7 Process Operating Hours Per Day

The data for process operating hours per day, figure (7), clearly show that most of the heat-activated processes within the sample frame are operated for 24 hours per day. A rather flat distribution is obtained for the remainder with the exception of a significant peak representing "single shift" operations.

3.2.8 Process Operation Hours Per Week

The data for operating hours per week, figure (8), are characteristic of the reduced working week currently in force in UK industry. It can be seen that the weekly (i.e. 7 days) continuous processes reduce to some 26% whereas about 50% of processes are continuous with respect to a 24 hour time scale.

4.0 Analysis of Derived Quantities

The capital pay-back period for a storage system is inversely proportional to the annual charge frequency. Given an industrial pay-back period constraint of less than 3 years, the target area for economic applications of storage is restricted to annual charge frequencies greater than 500. A charge frequency of 500 corresponds to an approximate process period of 8 hours and represents an upper economic constraint for most storage technologies.

Figure (9) shows the cumulative frequency distribution for operating cycle period length. It can be seen that approximately 50% of the cumulative frequency is characterised by continuous processes where a default period length of 1 minute has been assumed. About 75% of the survey results are within the economic constraint shown (including those processes classified as continuous). A further 20% of the cumulative frequency lies in the cycle period range 8 to 16 hours.

Although many processes were considered by the respondees to be continuous, it is probable that a significant number of sub-periods exists within this category. It is also probable that such sub-periods would not exceed 8 hours. Consequently a more detailed study of so called continuous processes would not significantly alter the above conclusion.

If it were possible to incorporate storage into all the processes specified in the data base and to operate such a system around the process operating temperature, only a fraction of the processes would be feasible. As the process periodicity is reduced, it becomes necessary to charge the store at higher powers. Thus a technical constraint, in addition to the above economic constraints, must be applied to the data base which excludes all those processes where the power density exceeds that which is feasible for a particular storage technology.

The target power density for a particular process can be defined by:-

$$\text{Power density} = \frac{\text{Energy Density}}{\text{Period}}$$

Therefore, with a knowledge of the process period and the energy density, the required power density can be evaluated. The results are shown in figures (10 to 14) which also indicate the limits due to the economic and technical constraints. The figures also give the percentage of processes that might use a particular storage technology and set upper limits to the feasible number of processes for storage applications.

4.1 Discussion of Storage Potential

4.1.1 Hot Water Storage Technology

The storage of heat in water is limited by the ability to store the heat effectively. The most effective method of storage is the stratified store, but more often the store is "fully mixed" by the charging procedure which reduces its effectiveness. Hence a limit has been set at the power density which will give a fully mixed condition in a vertical cylindrical type store of unity aspect ratio. This limit has been estimated to be 100 kW m^{-3}.

Applying the above limit, as shown in figure (10), only 0.37% of processes in the data base of temperatures less than 50°C reside within the feasible region. There are 2.7% of processes in the range $51^{o}C \leqslant T \leqslant 85^{o}C$. However, both hot water storage categories appear to be more significantly constrained by the economic periodicity limit of 8 hours.

4.1.2 Steam Storage Technology

Steam storage does not appear to have a definite technological limit for power density. It is uncertain whether or not the so-called continuous processes should be included in the percentages of processes in the feasible range, but it has been found that many continuous processes which use steam would benefit from the inclusion of an appropriately designed accumulator.

From figure (11) it can be seen that 23.8% of processes reside within the feasible region for $86^{o}C \leqslant T \leqslant 150^{o}C$ and 17.1% of processes in the range $151^{o}C \leqslant T \leqslant 250^{o}C$. Again the steam storage category is most significantly constrained by the economic periodicity limit of 8 hours.

4.1.3 Regenerator Storage Technology

The energy density will change between the two temperature bands which are considered to be suitable for regenerators: the lower temperature is more suited to feolite and the upper temperature band is more appropriate for fireclay.

The technological limit is based upon a glass industry regenerator raising the temperature of air by $1000^{o}C$. This gives a limit of about 1300 kW m^{-3}.

From figure (13) it can be seen that 2.8% of processes reside within the feasible region for the temperature range $501^{o}C \leqslant T \leqslant 999^{o}C$. Some 6.3% of processes reside within the range $T \geqslant 1000^{o}C$.

4.1.4 Phase-Change Storage Technology

There are five temperature bands which are considered suitable for latent heat storage, up to a maximum of $500^{o}C$. The energy densities within this group increase for the two higher temperature bands.

The technological limit for the power density is currently considered to be around 100 kW m^{-3} depending upon the surface area available for heat transfer. From figure (13) it can be seen that, for the temperature range between $50^{o}C$ and $250^{o}C$ the potential feasible number processes is smaller than that for the competitive water and steam storage technologies. Small but significant numbers (0.33%) of processes are feasible in the temperature range $251^{o}C \leqslant T \leqslant 500^{o}C$.

4.1.5 Thermo-chemical Storage Technology

Thermo-chemical storage is considered applicable only to the three lower temperature ranges with one level of energy density which is higher than most of the other systems. The power density though is thought to be around 100 kW m^{-3} and the economic limit is about 125 kW m^{-3}, which means that there are no processes in the feasible range as shown in figure (14).

4.2 Energy Potential Scaling for the UK

Figure (11) indicates that steam storage technology is by far the most significant for industrial applications.

In general, however, where heat transfer occurs across solid boundaries, the achievable power density is inversely proportional to energy density. Therefore, thermo-chemical storage technologies, although possessing high energy densities, when used in industrial processes, their final volume will be dominated by the heat exchanger size, thus reducing the advantage of the higher energy density storage of heat.

Figure (15) summarises the present analysis of storage technologies in the form of a histogram.

The most important part of this stage of the investigation was in establishing a methodology by which reasonably accurate estimates of the energy conservation potential could be made. The key to this methodology is in the recognition of the non-linear relationship between process energy use and process operating temperature. Most previous studies of industrial conservation potential have not treated this aspect in sufficient detail with the result that global figures for energy conservation potential have been put forward. Thus little has been known about where exactly the major emphasis of conservation should be applied.

Figure (16) shows the percentage of industrial process heat required at temperatures less than or equal to the process operating temperature (T). Also shown in figure (16) are two results from surveys conducted in the USA. The present results show a remarkable similarity to the USA result with the greatest deviation occurring at low temperature. This is probably associated with the poor response from sectors such as the food processing industry where low temperature process heat use is expected to be significant. It is interesting to note that for the sample frame, 50% of industrial process heat is required at temperatures less than 200°C.

Scaling the percentage of feasible processes residing within the economic and technological constraints by the reference energy used in that temperature range, results in the technical and economic potential of 9.8 x 10^{6}J (3.3 MTCE) for the sample frame as a whole.

The above survey result represent an upper limit which assumes that all application opportunities will be realised. However, if the concept of cascading between processes is introduced, then the upper limit can be refined if the probability of on-site cascade matches between processes is known.

4.2.1 Probability of On-Site Cascade Matching Between Processes

Individual survey sites were selected from the data base and arranged into a cross-tabulation of number of plants and operating temperature. The processes which matched in terms of operating temperature, were summated and the probability of two or more on-site cascade matches established for the sample frame as a function of operating temperature. Figure (17) shows the cumulative probability of on-site cascade matches between processes as a function of the process temperature. When these probabilities are used to weight the previous results, the total technical and economic potential becomes 4.4% of the 1980 reference energy use for the sample frame i.e. 3.9×10^{16}J (1.3 MTCE). This represents 1% of the total UK industrial energy consumption.

The results of the three levels of analysis are summarised in table (1).

4.3 Energy Conservation Potential - continuous processes

The highest priority for pure heat exchange is in waste heat recovery from continuous processes. Applying the previous methodology but excluding the storage and cascade probability constraints gives the continuous process waste heat recovery potential as 8.67×10^{16} J/yr (2.9 MTCE). This figure assumes that all the waste heat is recycled to the process and therefore clearly represents an upper limit to the technical potential for the sample frame.

4.4 Energy Conservation Potential - all processes

To compare the results of the survey with other UK data it has been necessary to assume that all processes in the data base are part of the global technological potential. For the whole sample frame the technological energy conservation potential amounts to 17.2×10^{16} J/yr (or 5.8 MTCE/yr).

The UK department of energy ref (4) has estimated that the technological energy conservation potential in the industrial sector is in the range 11 to 16 MTCE/yr out of a total expenditure of 130 MTCE/yr. In order to prepare an approximate comparison, the scale factor for the sample frame energy use compared with the total UK consumption, was assumed to apply equally to the results of the previous sections. If this is done, then the comparative technical potential from the survey is in the range 12.5 to 25 MTCE/yr. Although the lower limit is in good agreement with UK government estimates, the upper limit exceeds government predictions by 9 MTCE/yr.

5.0 Conclusions

There are many impediments to energy conservation in industry; attitudes, technological delays and economic bias, amongst others. However, the regions identified for the application of storage and in particular the technological potential of 1.3

MTCE/year resides within industrial sectors where conventional waste heat recovery will be, in many cases, an economically superior option to storage.

One of the major advantages of storage, however, is that waste-heat streams, for which no direct waste heat recovery application has been found, can be used either in plant or in across plant cascading. If the conditions for the probability of cascade matching apply generally in industry, then only continuous processes will be considered for waste heat recovery. It seems probable therefore that up to 60% of the technological potential for waste-heat recovery will not be implemented due to no suitable sink being available for the heat. Storage cascades could, therefore, account exclusively for only 0.8 MTCE/year within the sample frame.

Although the UK total industrial potential represents 3.41 MTCE/year, the bulk of the applications are bound to reside within the temperature ranges $86^{\circ}C \leqslant T \leqslant 250^{\circ}C$ and $T \geqslant 1000^{\circ}C$. These account for 97% of the potential. The lower of these temperature ranges is clearly associated with steam usage, and steam storage is therefore applicable: the higher range is an area where traditional regenerators have been used. For the latter temperature range, glass industry type regenerators could be expected to contribute to the conservation potential by 0.25 MTCE/year ref (4). This accounts for 41% of the technological potential highlighted by this survey.

The greatest potential for steam storage is not in association with an individual process, but integrated into the process steam service. Waste-heat recovery devoted towards steam raising which can then be integrated into the steam service system is likely to be the most profitable approach.

REFERENCES

1. '1968 Census of Production' HMSO

2. 'Digest of UK Energy Statistics 1981' Department of Energy HMSO

3. "S.P.S.S. Update 7-9" Hull and Nie, McGraw-Hill, U.S. 1981

4. "Energy Paper 32" Department of Energy, HMSO, 1978.

5. 'Turbonetics Energy Builds on R & D Base' Palmer, J.D., Turbomachinery International, Connecticut, USA, Jan/Feb 82.

6. 'Analysis of the Economic Potential of Solar Thermal Energy to Provide Industrial Process Heat in the USA', Czego and Frazer, M. Solar Energy for Industry Conference (C14), Royal Institution, Feb 1978.

DISCUSSION

Question - Dr Ehringer, CEC : 1. Did you also consider the economical aspects ?
2. Why is there no chance on the thermochemical energy storage ?

Answer 1. Yes. These are generally considered in the charge frequency constraint. Lower charge frequencies give longer payback periods. However, a frequency of 500 cycles/yr has a certain tolerance and is only therefore a guide.

2. Unless design power densities exceed 100 kW/m3 or the cycle frequency constraint is relaxed, i.e. towards long term storage with solar for instance, there is little potential for thermochemical storage.

STORAGE TECHNOLOGY		OPERATING TEMPERATURE (^oC) <50	51 - 85	86 - 150	151 - 250	251-500	501 - 999	>1000
HOT WATER STORAGE	% of Feasible Processes	0.37	2.70					
	% Global 1980* Ref. Use	0.012	0.22					
	% Cascade Match	0.001	0.05					
STEAM ACCUMULATORS	% of Feasible Processes			21.83	17.11			
	% Global 1980* Ref. Use			6.24	2.92			
	% Cascade Match			2.92	0.88			
REGENERATORS	% of Feasible Processes						2.77	6.30
	% Global 1980* Ref. Use						0.28	1.37
	% Cascade Match						0.07	0.47
LATENT HEAT STORAGE	% of Feasible Processes	0.30	2.50	7.67	5.02	0.33		
	% Global 1980* Ref. Use	0.01	0.20	2.19	0.86	0.024		
	% Cascade Match	-	-	-	-	-0.004		
THERMOCHEMICAL ENERGY STORAGE	% of Feasible Processes	0	0	0				
	% Global 1980* Ref. Use	-	-	-				
	% Cascade Match	-	-	-				

TABLE 1 RESULTS OF ANALYSIS OF THE ENERGY CONSERVATION POTENTIAL for storage technologies as a function of process operating temperature.

* 1980 Reference energy use for the sample frame = 89×10^{16} J/annum.

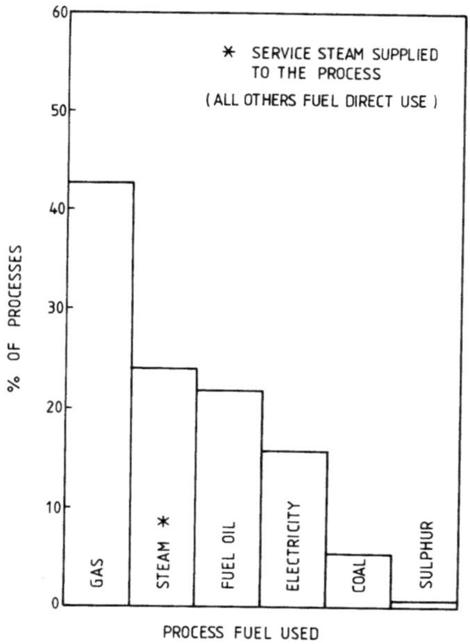

Fig. 1 Distribution of Process Fuels for the Survey

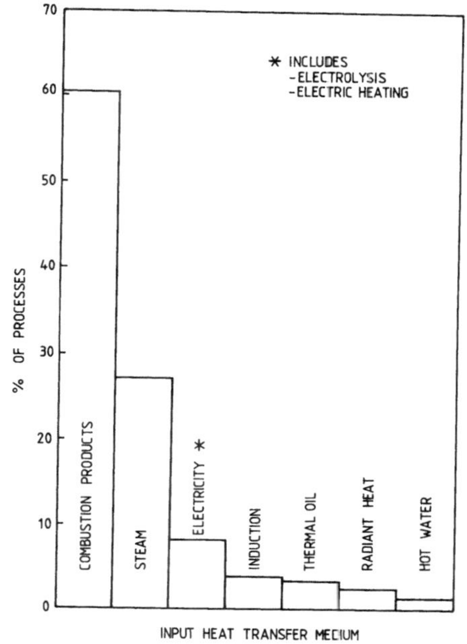

Fig. 2 Distribution of Input Heat Transfer Medium for the Survey

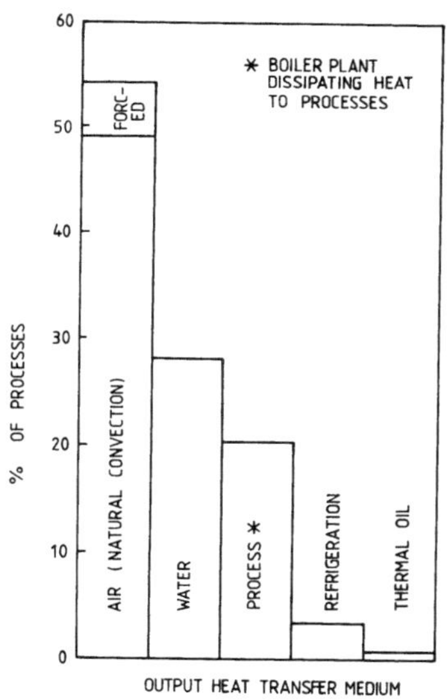

Fig. 3 Distribution of Output Heat Transfer Medium for the Survey

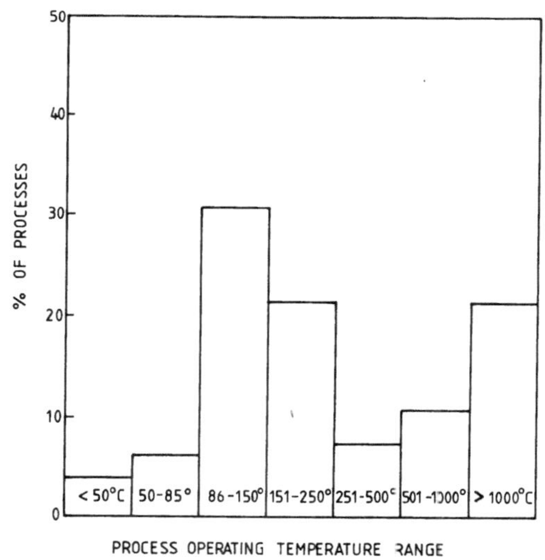

Fig. 4 Distribution of Process Operating Temperature Range for the Survey

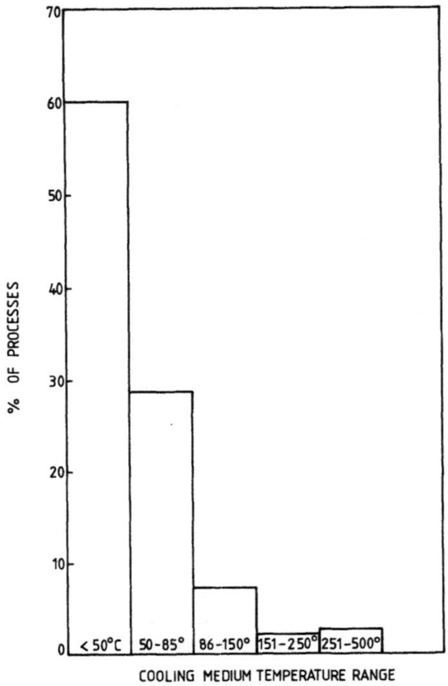

Fig. 5 Distribution of Cooling Medium Temperatures for the Survey

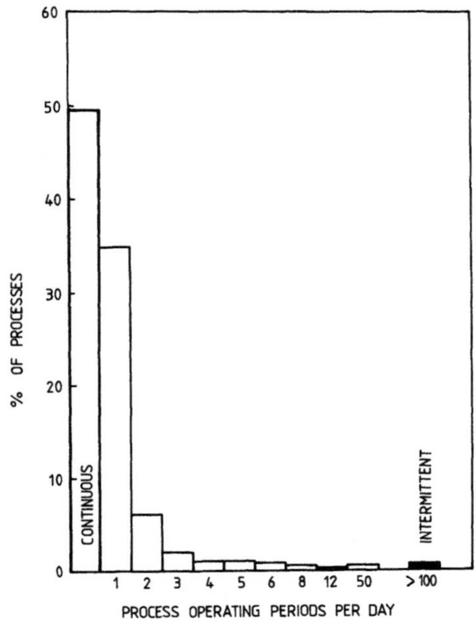

Fig. 6 Distribution of Process Operating Periods per Day

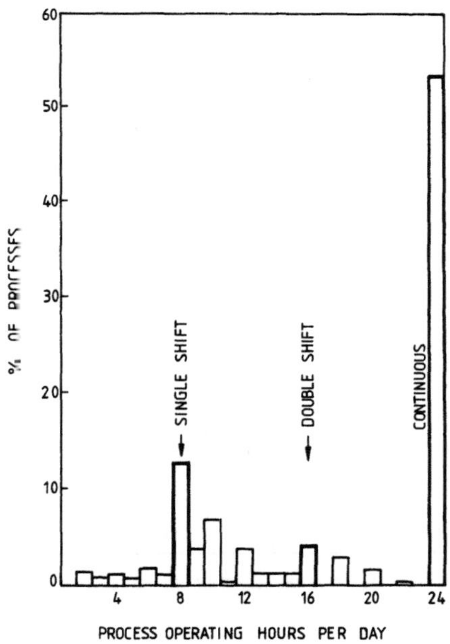

Fig. 7 Distribution of Process Operating Hours per Day

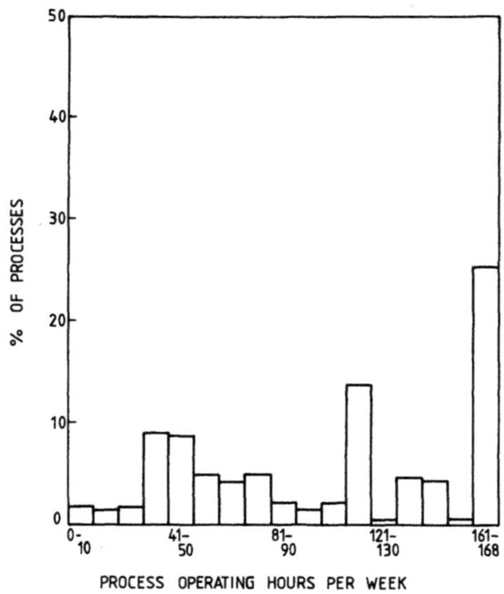

Fig. 8 Distribution of Process Operating Hours per Week

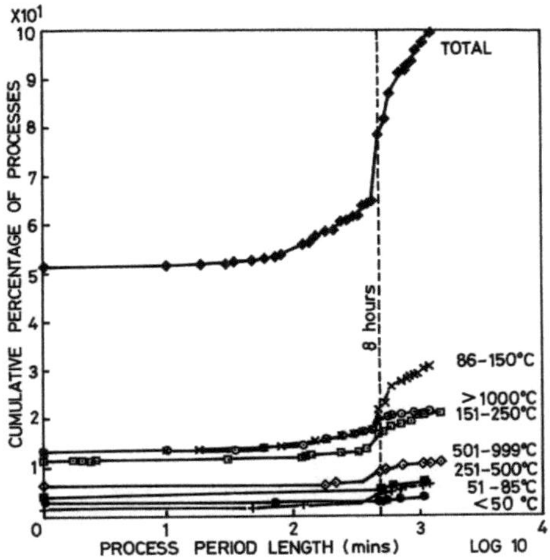

Fig. 9 Distribution of Period Length with Process Operating Temperature

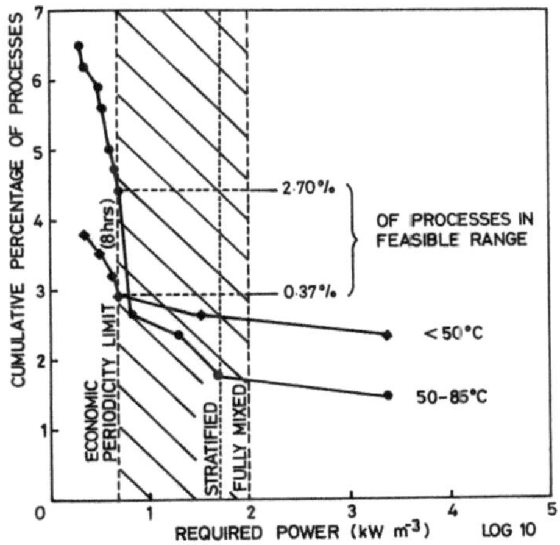

Fig. 10 Distribution of Power Density with Cumulative Percentage of Processes (Hot Water Storage)

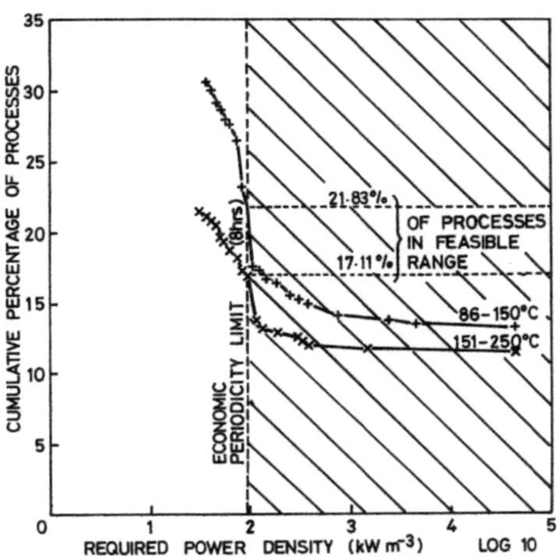

Fig. 11 Distribution of Power Density with Cumulative Percentage of Processes (Regenerator Technology)

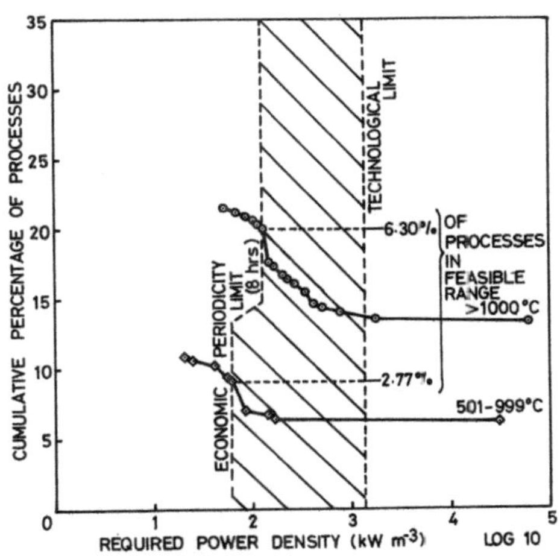

Fig. 12 Distribution of Power Density with Cumulative Percentage of Processes (Latent Heat Storage Technology)

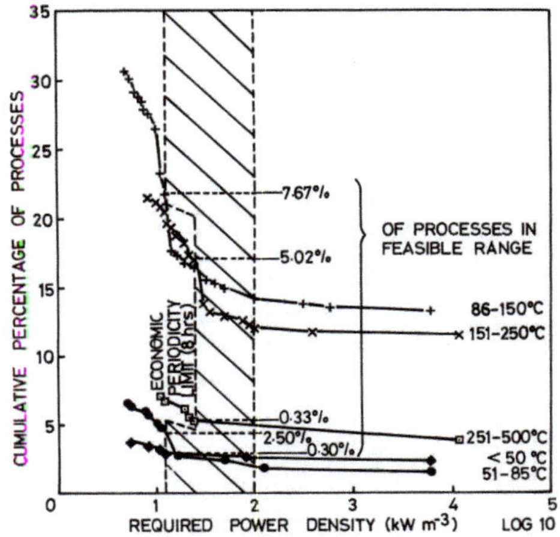

Fig. 13 Distribution of Power Density with Cumulative Percentage of Processes (Latent Heat Storage Technology)

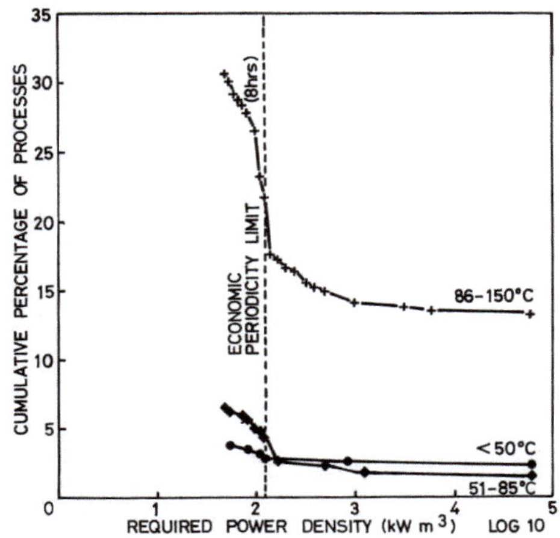

Fig. 14 Distribution of Power Density with Cumulative Percentage of Processes (Thermochemical Storage Technology)

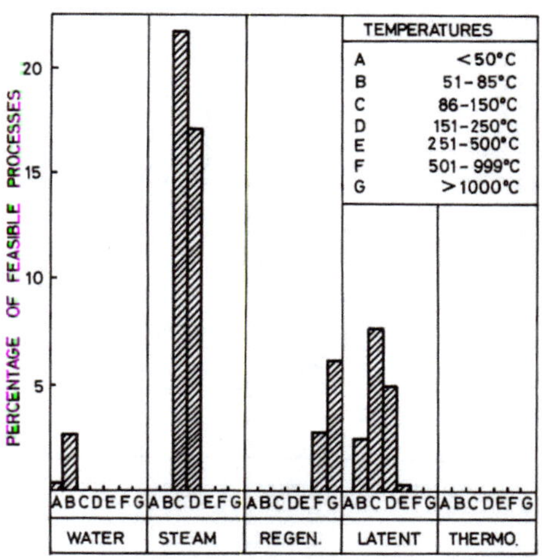

Fig. 15 Histogram of the Percentage of Processes feasible for Storage Technology

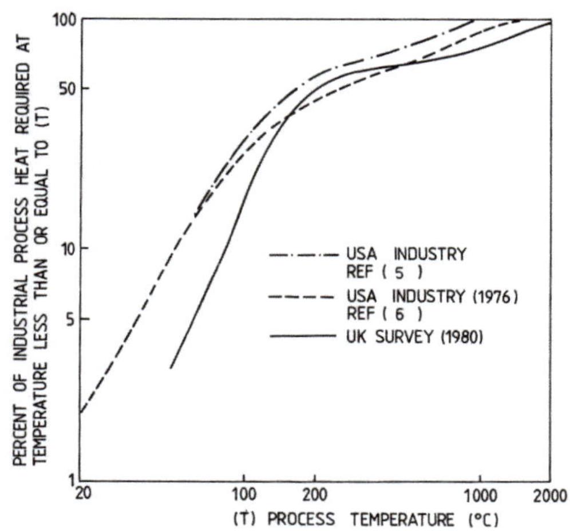

Fig. 16 Industrial Energy Consumption as a Function of Process Temperature (100% = 1980 Reference Energy Consumption for Sample Frame of $89x10^{16}$)

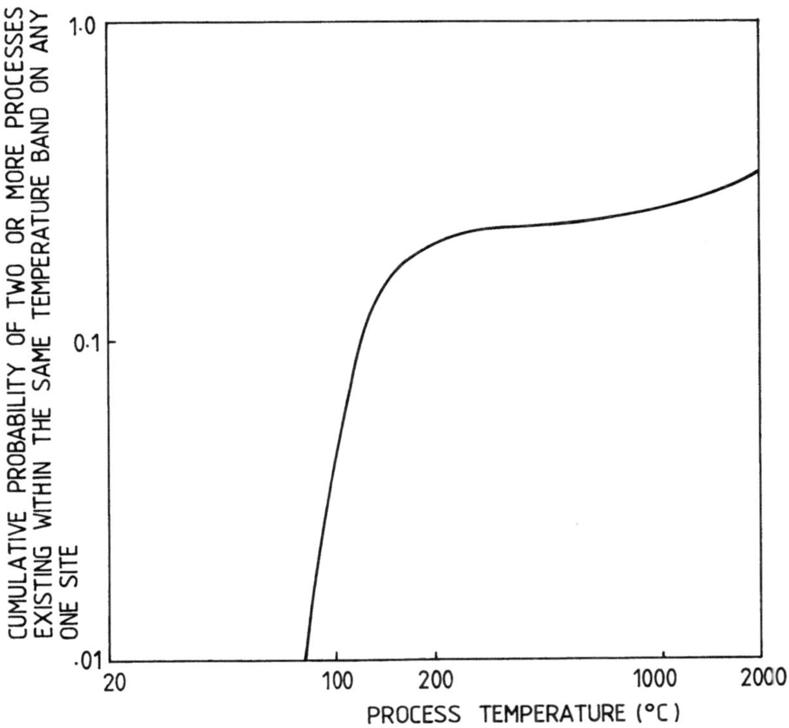

Fig. 17 Probability of "On-Site" Process Cascade Matching as a Function of Process Temperature

MIX
Papier aus verantwortungsvollen Quellen
Paper from responsible sources
FSC® C105338

If you have any concerns about our products,
you can contact us on
ProductSafety@springernature.com

In case Publisher is established outside the EU,
the EU authorized representative is:
Springer Nature Customer Service Center GmbH
Europaplatz 3, 69115 Heidelberg, Germany

Printed by Libri Plureos GmbH
in Hamburg, Germany